EBB AND FLOW

Ebb and Flow

THE TIDES OF EARTH, AIR, AND WATER

by Albert Defant

ANN ARBOR
THE UNIVERSITY OF MICHIGAN PRESS

Published in the United States of America by
The University of Michigan Press and simultaneously
in Rexdale, Canada, by Ambassador Books Limited

Library of Congress Catalog Card No. 58-62520

First published as Ebbe und Flut des Meeres der Atmosphäre und der Erdfeste, 1953 *by Springer-Verlag, Berlin–Göttingen–Heidelberg*

Translated by A. J. Pomerans

Designed by George Lenox

Manufactured in the United States of America

Contents

EBB AND FLOW

I. *What Are Tides?*

Watching from the Shore

The tides are the heartbeat of the ocean, a pulse that can be felt all over the world. This mysterious rhythm has gripped man's imagination since the beginnings of recorded history, spurring him to try to understand its causes and grasp its meaning.

An attentive observer, watching the ebb and flow of the tides from a vantage point on the coast, will notice that this movement of the water includes both a vertical and a horizontal motion. Both recur at equal intervals, i.e. they have the same period, and both are part of the same wave motion. No coastline is without tides—the periodic rise and fall of the water level is universal. Often it is so weak that the effects of wind and weather obscure it, but equally often it presents us with the magnificent spectacle of tides of up to sixty feet. These tremendous fluctuations have a marked effect on the economy and way of life of all those who live near the sea.

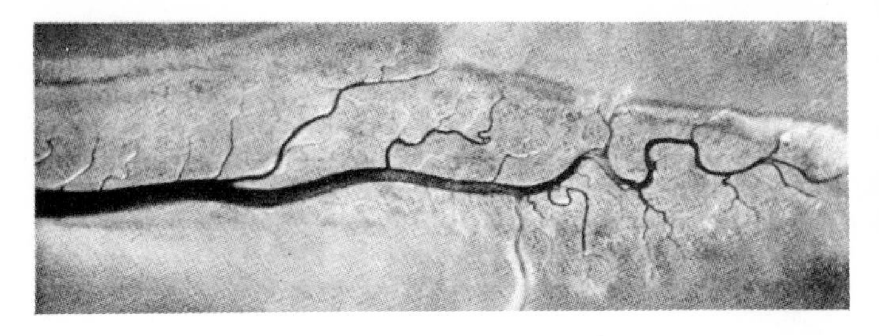

FIG. 1. Mud flats during low tide, showing channel and tributaries.

Tides are particularly impressive in shallow waters. Off the German North Sea coast, for instance, a strip from some six to some twelve miles broad is above water during low tide and submerged during high. This coastal strip is a wide, barren, almost level, gray surface, composed of mud and wet sand and known as the German Watt, an amphibian belt which is sea bed and mainland at different times. Behind these mud flats are low-lying, fertile marshes, protected for the most part by dykes. During high water the dykes form the barrier between land and sea, and tremendous catastrophes occur whenever they succumb to the pressure of the water. The extensive mud flats are divided by clearly defined channels (Fig. 1). During low tide the channels drain the water to the sea, in much the same way as would a system of rivers, and during high tide they lead it to the innermost parts of the flats. The larger and deeper of these channels, which are navigable, are marked off by poles from the surrounding shallows. From these channels the water spreads over the entire area.

When the coast is steep, the picture is of course quite different. Here the tides appear almost exclusively as a rise in the water level, while the horizontal displacement has become negligible. Instead of submerging large tracts of land, the tidal effect is here restricted to a narrow coastal belt.

Tides at the Coast and on the Open Sea

We have seen that tides recur with great regularity and uniformity. Now we shall briefly discuss the basic phenomena and introduce a few technical terms. With the approach of the crest of the tidal wave, the water level rises; this is called the flood tide, culminating in high water or high tide. The water level then begins to drop, the tide turns; this is called the ebb tide. Once the water has reached its lowest point we have low water or low tide, and now the unceasing cycle starts anew. More detailed observation of this regular and recurring phenomenon reveals that the mean interval between two successive high, or low, waters is 12 hours, 25 minutes; if we have high water at 8 a.m. today, the next high water will take place 12 hours, 25 minutes later (at 8:25 p.m.), and the next in another 12 hours, 25 minutes (at 8:50 a.m. tomorrow). Thus each day the tides rise on an average 50 minutes later.

We reckon our time by the apparent motion of the sun. The sun takes 24 solar hours to make one apparent revolution round the earth. At noon it passes through a given meridian and after a full revolution, for which it needs 24 mean solar hours, it passes through the same meridian again. The moon, on the other hand, passes through the meridian at mean intervals of 24 hours, 50 minutes; we call this interval a lunar day. Thus the moon passes a particular meridian 50 minutes later each day, which corresponds with the average lag in the tides. The interval between the passing of the moon and the rise of the tides is constant for any given coastal position; every harbor has its characteristic tidal establishment or lunitidal interval. For instance, when we say that Heligoland has an establishment of 11 hours, 20 minutes, we mean that, with minor deviations, high water in Heligoland always occurs 11 hours, 20 minutes after the moon has passed the local meridian.

The position of the moon affects not only the time of the tides but also the height and the mass of water involved in the tidal current. During full moon and new moon (syzygies) the difference between high and low water, known as the range or amplitude of the tides, is at its maximum—we have the highest high waters and the lowest low waters. These are called spring tides. During the first and last quarters of the moon (quadratures) the range is particularly small, and we have neap tides. This difference in the range is known as the phase inequality. It, too, depends on the relative positions of the moon and the earth, but it is also affected by the position of the sun, since the phases of the moon depend on the position of the earth relative to both these celestial bodies. The tidal range sometimes varies from tide to tide; the morning high water may be higher than the subsequent afternoon high water, or vice versa. This phenomenon is known as the diurnal (daily) inequality. There are other inequalities, but the two we have mentioned are the basic ones.

The periodic nature of tides has been known to sailors since antiquity. To understand the reason why the ancient Greeks, otherwise such good observers, hardly mentioned tides, we need only remember that the Mediterranean, the world of the Greeks, is practically tideless. Those ancients who did reach coastal regions open to the oceans were all the more struck by the effect. Many men with keen intellects (Herodotus, Pytheas, and Aristotle among the Greeks, and Posidonius and Strabo among the Romans) were familiar with the tides and showed a great deal of insight in discussing them. On the other hand, Alexander the Great and Julius Caesar were taken quite unawares (the former at the mouth of the Indus and the latter on the shores of Great Britain) because of their ignorance of local tides. These stories are

FIG. 2. High and low spring tides in the Bay of Fundy (New Brunswick, Canada).

dramatically reported by Curtus Rufus in his biography of Alexander the Great and by Caesar himself in his *Gallic Wars.*

The range of the tides differs from one coastal town to the next; the rise varies from a few inches to sixty feet and more. The tides in the Bay of Fundy, New Brunswick, are famous for their range of almost seventy feet on the southern side of the bay. Figure 2 shows a landing stage in the Bay of Fundy during low water (top) and high water (bottom). The water has darkened the logs (top), and we can gauge the height of the high-water mark by comparing the logs with the horse and the people in the lower picture. In Europe the tidal range is particularly great at Mont-Saint-Michel, off the coast of Normandy. This small island lies in mud flats that are some forty feet below water during high spring tides, and some twenty feet below during high neap tides, but it can always be reached from the mainland during low water. Most of the world's great ports are on tidal rivers near shallow coastal waters where the tidal range is usually considerable. Here, all harbor traffic has to wait for the tides. Everyone who has approached Hamburg by sea has noticed the great number of ships awaiting the flood tide before sailing up the Elbe to the harbor. Experienced pilots, familiar with every action of the tides in the estuary, are needed to guide ships safely to and from the harbor.

Since the velocity of tidal currents can be very great, particularly near the coast and continental shelves, slow-going ships are often greatly affected. In such waters knowledge of the direction and magnitude of the tidal currents is essential, since unfavorable tidal currents may delay ships by many hours. On the open sea or in deep water the tidal current is weak and correspondingly less effective.

On the high sea, where there is no fixed plane of reference for comparing changes in sea level, it is very difficult to measure the tides. In the case of a ship at anchor, however, the oceanic tidal currents result in a measurable drag on the cable; and in shallow waters (up to approximately 300 feet), one can estimate the tidal range by sounding the depth. Measurements of this kind are, of course, impossible when the ship is moving.

Wind and the Tides

The regular course of the tides is often altered by winds. Breezes from seaward drive the water towards the coast, and the water level rises; when the breeze is from the land the water level drops. In both cases there are corresponding changes in barometric pressure. These variations are independent of the tides and even occur on expanses of water, such as inland lakes or the Baltic Sea, that have little or no tide. Coupled with tides, particularly with high spring tides, they may become extremely dangerous. If wind and tide reach their peaks together we have tidal surges which may breach dykes and lead to catastrophic losses of life and property. Figure 3 shows a marigram (a record of tidal fluctuations measured by a tide gauge) of a tidal surge in

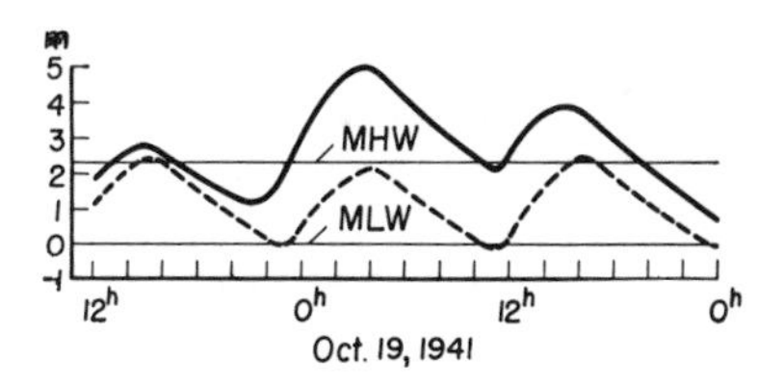

FIG. 3. Tidal surge at Hamburg on October 19, 1941. (-------- expected tides; ______ recorded surge). MHW: Mean High Water; MLW: Mean Low Water.

Hamburg during October 18–19, 1941. The dotted line shows the normal tidal fluctuations of the water level, the unbroken line the fluctuations actually recorded. The sea level had been considerably raised by the effects of wind and air pressure. Tidal surges have caused many changes in the world's coastlines, particularly in places with shallow waters. There is hardly a part of the German North Sea coast that has retained its shape over the ages. In the course of time the sea has encroached upon enormous tracts of land, and only by erecting artificial barriers can man halt this relentless process (Figs. 4, 5, and 6). The protection of islands and coasts against these destructive forces of nature is an important task that calls for long-range planning.

FIG. 4. Norderhafen on Nordstrand (northern Germany); tidal surge on October 18, 1936. The house, built on the dyke itself, was severely damaged. (Photograph: A. Busch, Morsumhafen.)

FIG. 5. Destruction of the Nordstrand Dyke after successive tidal surges on October 27, 1936. (Photograph: A. Busch, Morsumhafen.)

FIG. 6. Near Nordhafen on Nordstrand (seen from the roof of a stables on the marsh) on October 18, 1936. The high water has reached the dyke and is pouring over the top. (Photograph: A. Busch, Morsumhafen.)

II. *Measuring the Tides*

Tide Gauges

To make a close study of the tides at a particular point, we shall have to take readings every hour over a large period of time. On shallow coasts, where there is little surf, or wherever surface disturbances of the water level are minimal, the tidal range can be measured by a series of graduated poles driven into the sea bed at intervals (Fig. 7). The uppermost pole must reach above the water even during high water, and the lowest must be at such a distance as to give readings for the lowest low water. The graduations on the poles must, of course, be accurately levelled. Dams, landing stages, and other structures can also give a fair measure of tidal fluctuations. But all these devices are affected by surface waves.

If greater accuracy is needed, measurements must be made in such a way as to avoid the effect of waves. Near the shore, but above the high water mark, we sink a tank whose bcttom is some three to six feet below the level of lowest low water (Fig. 8). The tank is connected to the ocean by a narrow pipe which drops down to a considerable depth. The end of the pipe is covered with a device like the sprinkler on a watering can, and a buoy keeps it off the sea bed. Since the effect of the waves is not felt at this depth, the level of the water in the reservoir reflects only the movement of the tides. While the ocean itself shows a restless surface, that of

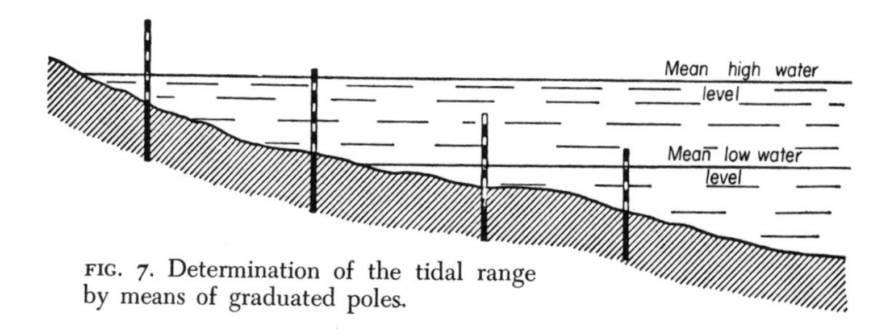

FIG. 7. Determination of the tidal range by means of graduated poles.

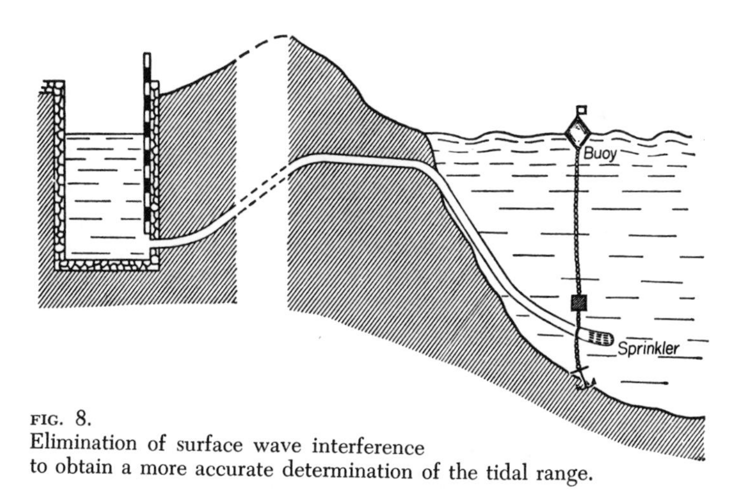

FIG. 8.
Elimination of surface wave interference to obtain a more accurate determination of the tidal range.

the reservoir is smooth, and the level can be measured by a graduated pole.

Measuring the tides by eye is laborious and time-wasting; for this reason we use a marigraph (self-registering tide gauge, such as was used to obtain Fig. 3) which gives a constant record of the water level. A float *F* (Fig. 9) rises and sinks with the water in the reservoir, *R*. A copper wire connected to this float passes over a drum, *G*, causing the drum to move with each change in the water level. The movement of the drum is transmitted to a stylus, S, which records the changes on a continuous strip of paper. All marigraphs are built on roughly the same principle, the details of which we need not discuss, apart from mentioning that the graph can be relayed electrically to the office that collects such data. The tidal curves obtained from a marigraph (Fig. 10) show the time and range of the tides at any moment. At a glance, they illustrate all the characteristics we have mentioned; there are two high waters per day, the tides arrive roughly 50 minutes late each day, and there are diurnal (daily) and phase inequalities.

Investigators must analyze all the data obtained, and determine for as many coastal points as possible the period of the tide, the amplitude of the tide, and the form of the tidal curve itself. Long-term observations at every point on a coast are, of course, out of the question, but we do have long-term data for the most important ports. From these we can derive approximations for any particular point nearby, provided that

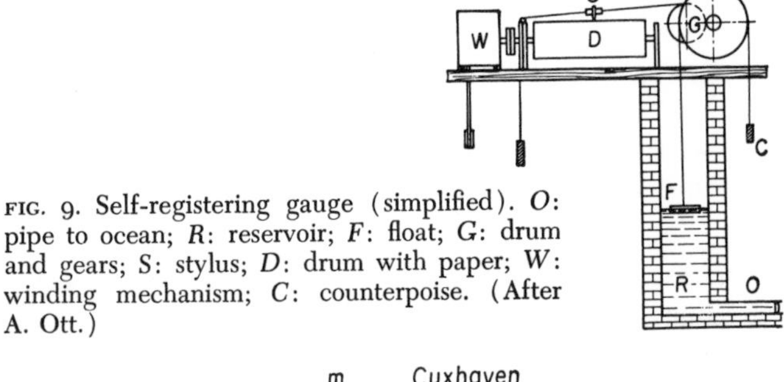

FIG. 9. Self-registering gauge (simplified). *O*: pipe to ocean; *R*: reservoir; *F*: float; *G*: drum and gears; S: stylus; *D*: drum with paper; *W*: winding mechanism; *C*: counterpoise. (After A. Ott.)

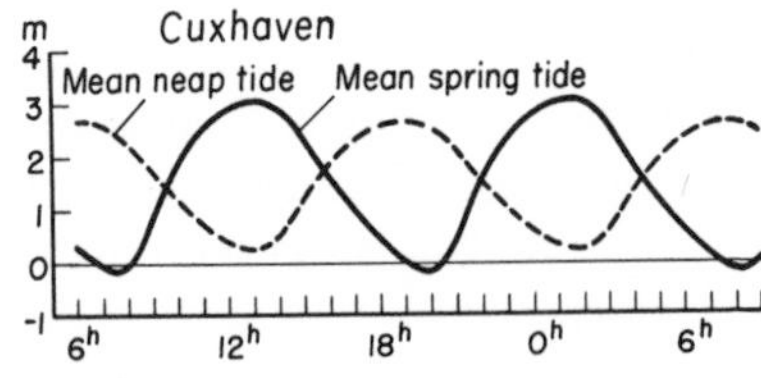

FIG. 10. Mean spring and neap tides at Cuxhaven. (Sketch by D.H.I.)

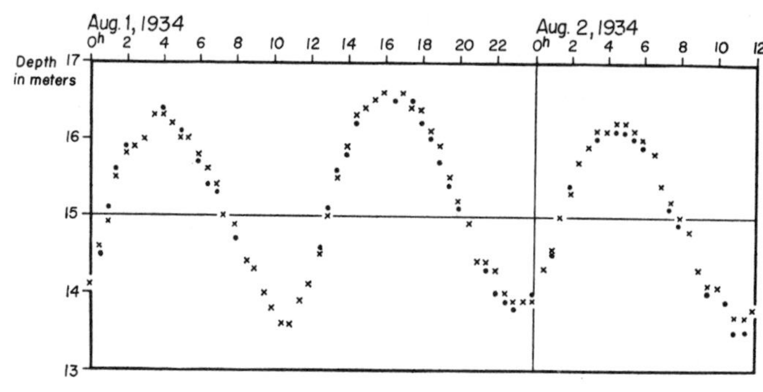

FIG. 11. Tidal curve from soundings off a ship at anchor; "Geeste-Station II" at 54°14′ N, 8°14′ E, North Sea.

we have simultaneous short-term observations for both the point and the port. We determine the period of high or low waters and the range, and correlate the figures so obtained with the long-term data for the port. This reduction method is successful because disturbances of the tidal cycle, particularly those caused by barometric pressure and wind, do not vary a great deal within a small area, and what small errors there are usually cancel out. Short-term measurements can in this way play a useful part in the evaluation of tidal phenomena over a fairly large coastal area, as well as in the compilation of tide tables.

Tides on the Open Sea

Tides cause fluctuations of the water level even on the open ocean. These changes are difficult to measure because the ship from which observations are made rises and falls with the water level, so that there is no fixed plane of reference. We have seen that in shallow waters up to some 300 feet deep the tidal range can be determined by systematic soundings from a ship at anchor. Figure 11 shows a curve thus obtained. Needless to say, the accuracy of such measurements is affected by wind and currents and, to a lesser extent, by barometric changes. Winds and currents can change the position of the ship if the anchor cable is slack, and if the sea bed is uneven the soundings will show variations in the water level that have nothing to do with the tides.

Pressure gauges have recently been devised to register the water pressure and thus the head of water above the instrument. Even with these gauges there are many good reasons why the water must not be deeper than some 600 feet, so that this method too is only applicable to shallow waters. The instruments are usually left in position for fourteen days; by distributing them over a given area we obtain a picture of the tidal changes in the area. Figure 12 shows a record obtained by this method. The unavoidable interference of wind, air pressure, and even silting must be averaged out, making this a cumbersome procedure. In deep waters on the open ocean the difficulties are so much increased that, even though the tidal range itself is fairly small, it is difficult to devise any accurate method of assessing it. However, an exact knowledge of the tides in deep waters is unnecessary, since here it makes little difference to navigators whether the depth is a few feet more or less.

Tidal Currents

Even in deep waters, however, it is important to know the direction and magnitude of the tidal current, since the current may considerably affect the course and speed of a ship. The current velocity must be determined from

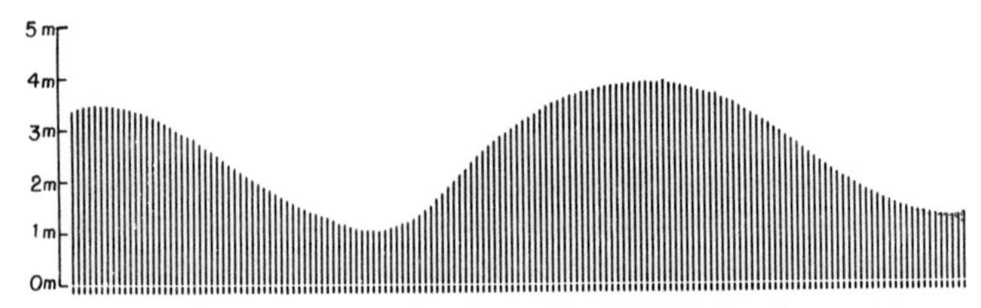

FIG. 12. Section from the records of a pressure gauge. (Photograph: D.H.I.)

FIG. 13. Current meter on board ship before being lowered into the water. (Photograph: Dr. G. Dietrich.)

a ship at anchor. The simplest device is the "log," a triangular board weighted at one edge so that it floats upright in the water. Short cords, fastened to its three corners, are joined to a line which is calibrated with knots. The log is launched from the ship. The amount of line run out during a measured interval of time indicates the speed of the surface current in knots (sea miles per hour). If we wish to determine the direction and magnitude of underwater currents we must use a current meter. Although there are a great many types of current meter, observers in recent years have been using one that works on the principle of the propeller. The instrument is lowered to the required depth, where it records the direction and magnitude of the tidal current.

Figure 13 shows the current meter on board ship. It has a streamlined body, from the top of which protrudes the actual measuring instrument. The latter is an airtight cylinder, approximately 18 inches in diameter, to whose casing twelve concave blades are affixed. This "propeller" has a specific gravity approximating that of

FIG. 14. Section of a current meter record during ten minutes. Left: number of revolutions of the blades. Right: compass readings; the pointer shows the direction of the current. (Magnified; width of original film 16 mm.)

water, and this enables it to turn on its bearings with the minimum of friction. The number of revolutions—and thus the magnitude of the current—is recorded photographically inside the cylinder. The current meter is kept parallel to the current by means of rear fins. Readings of a built-in compass are recorded on the same film. Photographs taken at intervals of five minutes give a continual record of the magnitude and direction of the current. Figure 14 shows a section from this type of photographic record. On the left is the record of revolutions over a period of ten minutes. At the beginning of the first five minutes the reading is 3777.6 (bottom). After the first five minutes it is 3800.1 (center), and after another five minutes it is 3819.3 (top). Thus the blades made 22.5 revolutions during the first five minutes, and 19.2 during the second. For the particular current meter, these figures correspond to a velocity of 24.9 and 21.9 cm/sec (i.e. 9.8 inches per sec, and 8.6 inches per sec) respectively. The compass readings are shown on the right. They were 243°, 249°, and 253° for the given period, where 0° means North, 90° East, 180° South, and 270° West. Clearly, a fairly constant current to WSW prevailed throughout the period.

Measurements of oceanic tidal currents are extremely difficult to make, since all readings are affected by the movement of the anchored ship and since there is no fixed plane of reference. Furthermore, tidal currents, like all ocean currents, show turbulence—short-term fluctuations in direction and magnitude—which can only be compensated by a long series of measurements.

As a rule the tidal current "turns" after every half-period (approximately every six-and-a-half hours). From maximum velocity it drops to zero, then rises to maximum velocity in the opposite direction; at the end of the half-period the turn is repeated. We call this a rectilinear (alternating) current, since flood and ebb currents have equal and opposite directions. More frequently, when it turns, the tidal current swings round to the right or left before finally changing its direction and increasing in strength. This is called a rotatory current. Figure 15 shows the direction and magnitude of the tidal current for a point 4.5 sea miles WSW from the lightship "Elbe I," from six hours before until six hours after high water in Heligoland. We can see that this is a current rotating in a clockwise direction with maximum velocity of roughly 1.4 sea miles per hour ESE or W respectively, and that the current turns at 6^h or 0^h respectively after high water in Heligoland.

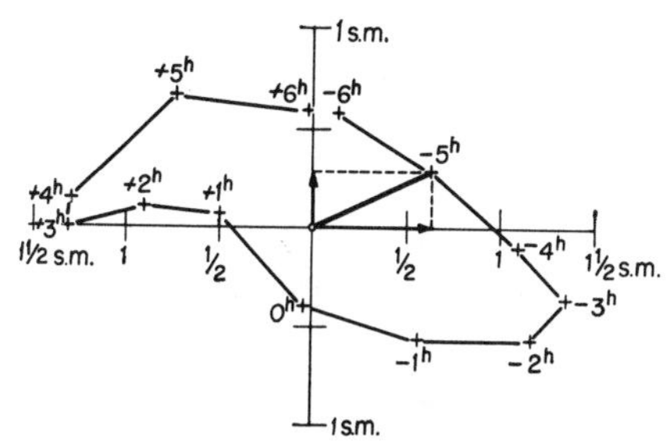

FIG. 15. Direction and magnitude of tidal current compiled from hourly observations 4.5 sea miles WSW from the lightship "Elbe I," from six hours before until six hours after high water in Heligoland (velocities in sea miles per hour).

III. *The Tide-generating Forces*

Effects of Moon and Sun

While we have very accurate knowledge of the magnitude, the distribution, and the period of the tide-generating forces themselves, our knowledge of their effects on the oceans, the atmosphere, and the land mass is rather limited. The basic facts alone are known and only developments in instrumental techniques and a long series of systematic observations can lead to further progress in this field.

To explain the tide-generating forces of moon and sun we must refer to Newton's law of gravitation. This law states that the force of attraction between two bodies is proportional to the product of their masses and inversely proportional to the square of their distance apart. It is this force of attraction which determines the annual motion of the earth about the sun (or to be precise, about their common center of gravity), and also the monthly motion of the moon about the earth (or rather about their common center of gravity). If earth and moon were not in motion, the force of attraction would cause a collision. However, their motion about the common center of gravity, like all circular motion, produces centrifugal forces which just balance the force of attraction. The centrifugal and attractive forces are equal and opposite to each other and act at the center of the earth or of the moon (Fig. 16). It is on this kind of equilibrium between opposing forces that the whole stability of the universe rests.

Though the earth-moon system as a whole is in equilibrium, individual particles on the surface of the earth are not. While the force of attraction on a particular particle depends on its distance from the moon, the centrifugal force is the same all over the surface of the earth, since every point on it describes the same motion about the center of gravity of the system. The resulting residual forces are such that their sum over all points on the surface of the earth is zero. These residual forces are the tide-generating forces. In Figure 17 the moon is directly overhead at Z (at its zenith), and here the force of attraction is greater than the centrifugal force. The difference is a small residual force directed towards the moon. At *N*, the point farthest away from the moon (i.e. where the moon is at its nadir), the force of attraction is smaller than the centrifugal force so that the residual force is directed away from the moon. In both cases the tide-generating force acts in a direction away from the center of the earth. Vectors for the two forces can

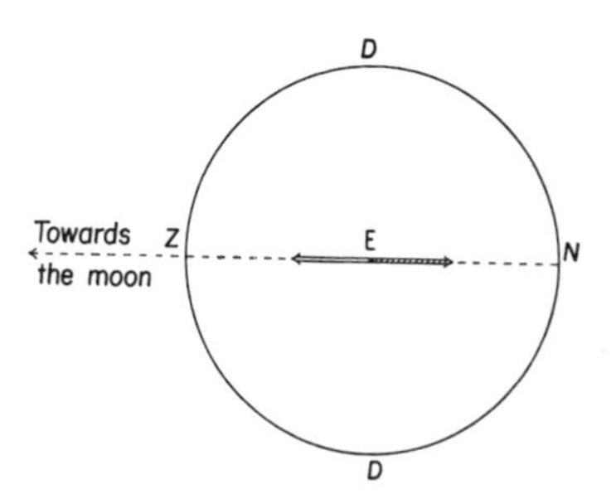

FIG. 16. Equilibrium between the total attractive force of the moon and the total centrifugal force.

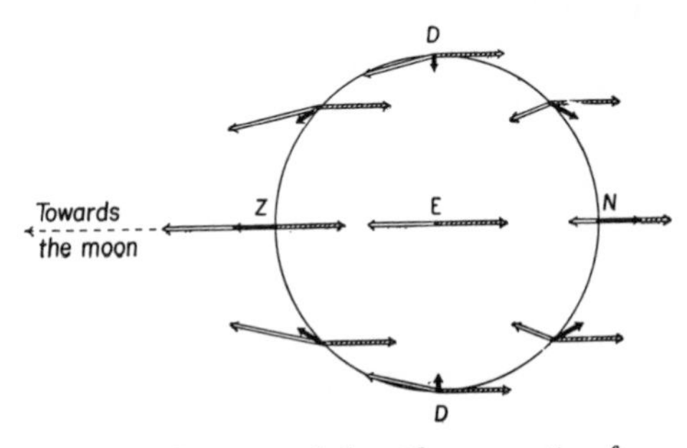

FIG. 17. The magnitude and direction of the tide-generating force as the difference between attractive and centrifugal forces at given points on the surface of the earth. Plain arrows: attractive force; striped arrows: centrifugal force; black arrows: tide-generating force.

be drawn for every point on the surface of the earth, and the resultants (shown on Fig. 17) form the system of tide-generating forces. Let us calculate the tide-generating force at the points *E*, Z, and *N*. At the center of the earth *E* (see Fig. 16), the force of attraction draws each kilogram of the earth's mass towards the moon with the very small force of +3.38 mg weight. The centrifugal force due to the earth's motion about the system's center of gravity is equal and opposite, i.e. —3.38 mg weight. Thus their sum is zero. At Z, which is nearer the moon, the force of attraction is somewhat greater (+3.49 mg), while at *N*, which is at a greater distance, the force is only +3.27 mg. Now at all three points (*E*, Z, and *N*) the centrifugal force is —3.38 mg. Thus the tide-generating force at Z is $+ 3.49 - 3.38 = + 0.11$ mg weight, while at *N* it is $+ 3.27 - 3.38 = - 0.11$ mg weight. Both forces are equal and opposite: at Z towards the moon, at *N* away from the moon; at both points the tide-generating force pulls "upwards," that is, away from the center *E* of the earth. For other points on the surface of the earth, the tide-generating force is no longer vertical. Figure 18 shows its distribution along a meridian. Only at the

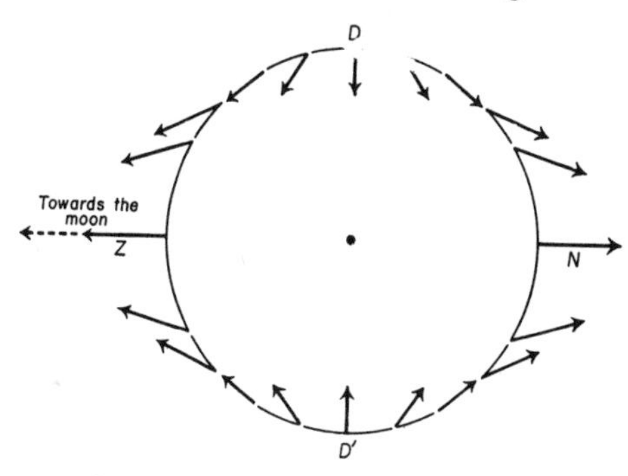

FIG. 18. Distribution of the tide-generating forces along a meridian.

four points *Z*, *N*, *D*, and *D′* does the tide-generating force act perpendicularly to the surface of the earth, i.e. straight up or down. Everywhere else it has a horizontal component. The arrows in Figure 18 have been greatly exaggerated. In order to get some idea of their real magnitude we must remember that where the tide-generating force acts upwards (at the points Z and *N* of Fig. 17), it will cause a man weighing ninety kilograms to lose only ten mg in weight, corresponding, say, to the weight of a single drop of sweat. The tide-generating force can increase or reduce gravity by a maximum of one nine-millionth, and compared with gravity, it can generally be neglected.

However, except at the points mentioned, the tide-generating force also has a horizontal component acting along the surface of the earth. This component will deflect a plumb line, a pendulum of length 12 m being deflected by roughly $1/1000$ mm. But despite its smallness, this horizontal component (the tractive force) is more important than the vertical component, since the other horizontal forces acting at the surface of the earth are usually of the same order of magnitude. In Figure 19, the horizontal component of the tide-generating force is zero at Z and *N*, and strongest along two circles at 45° to Z and *N*. In one hemisphere the force is always directed towards Z, in the other towards *N*, and on the great circle plane passing through the earth's center and perpendicular to the line Z*N* it also drops to zero. While all the vectors converge on Z in one hemisphere, they converge on *N* in the other.

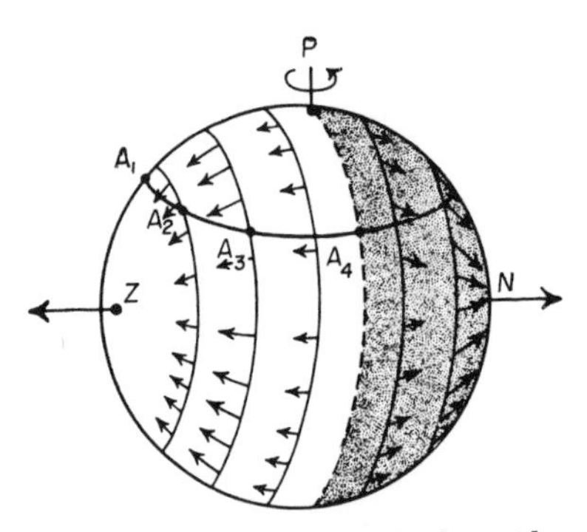

FIG. 19. The system of horizontal components of the tide-generating force on the surface of the earth. At Z the moon is at its zenith.

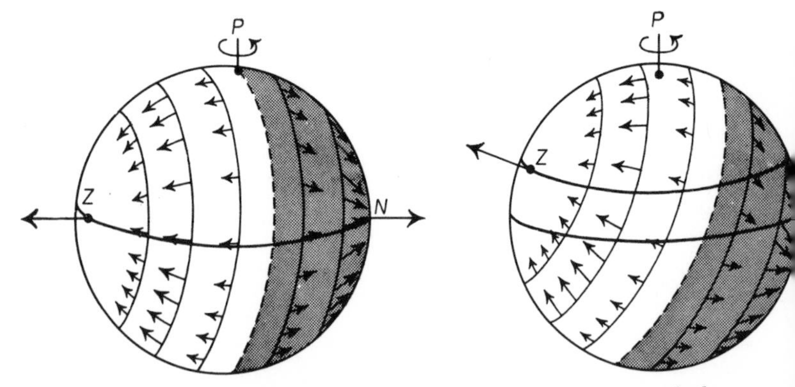

FIG. 20. Left: Distribution of horizontal components of the tide-generating force when the moon is in the plane of the earth's equator; right: when it is at its zenith over 28° N.

This system of tide-generating forces is determined by the position of the moon. As the moon changes its position, so the system of forces shown in Figure 19 moves with it. Figure 20 (left) shows what happens to this system when the moon is in the plane of the earth's equator and (right) when it is at 28° N.

Now the earth itself is not at rest but rotates about its own axis once a day. Let an observer be at the point A_1 (Fig. 19) at a given moment when the tide-generating force is directed towards the south. The earth's rotation will slowly bring him to the point A_2. The tractive force will increase and continue to do so until three hours later he reaches A_3. Here the tractive force is at its maximum. Six hours later, he has reached A_4 and the force becomes zero as he sees the moon setting. Then the force changes direction, reaching another maximum after a further three hours, and so on, until after twenty-four hours the observer will have returned to A_1. Thus the earth's rotation produces semidiurnal changes in the tide-generating force both in direction and in magnitude.

In Figure 21 the semidiurnal tidal effect of the moon is represented by force vectors from a central point.

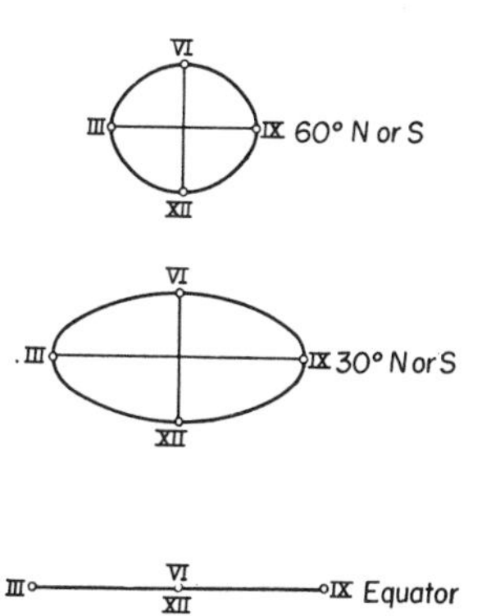

FIG. 21. The moon's semidiurnal tidal effect on a point in the Northern Hemisphere. The tidal force is represented in magnitude and direction by a force vector drawn from the center of the ellipse to the points of the clock.

At 0° (at the Equator), the tide is rectilinear (alternating), being zero at 0, 6, and 12 lunar hours, with a maximum to the east at 9 hours and one to the west at 3 hours; there is no north-south component. In other latitudes (e.g. at 30° or 60°) the force vector describes an ellipse. At 0 and 12 lunar hours it is directed towards the south, at 6 towards the north, at 3 towards the west and at 9 towards the east. Here the north-south component is of the same order of magnitude as the west-east component.

The tide-generating force of the moon must necessarily vary with the moon's distance from the earth. Now this distance changes continuously, if very slightly, in the course of one month, and the tide-generating force is affected by these small alterations. Furthermore the moon keeps altering its declination and the resulting asymmetry of the system of tide-generating forces with respect to the earth's equator (see Fig. 21) causes the diurnal inequality in the tide-generating force. The greater the declination, the greater this inequality.

So far we have considered the moon, but clearly the sun too produces a system of tide-generating forces. The semidiurnal tidal effect of the sun is only about one half as great as that of the moon. This is due to the fact that, while the moon has a very much smaller mass than the sun, it is so very much closer to the earth that its proximity is the decisive factor. (If the mass of the sun were the same as that of the moon, the sun's tide-generating effect would be negligible.)

The fortnightly inequality is the result of the joint effects of moon and sun. During new moon, sun, moon, and earth are in one straight line. The moon's tidal effect is then added to the solar effect and the resultant tractive force is increased in the ratio 3 : 2, since the tide-generating force of the sun is half that of the moon. The same happens during full moon, when moon and sun are in opposition, i.e. above Z and *N* respectively (see Fig.

17). During first and last quarters, the moon has moved 90° away from the sun, and the resultant tractive force is roughly one half of the lunar force alone. This explains the occurrence of spring and neap tides.

The other planets in our solar system are too small and too far from the earth to have any noticeable tide-generating effect. Thus the tide-generating effects of moon and sun alone need be considered, and these have been computed with great accuracy. There is little difficulty in calculating the magnitude and direction of their resultant tractive forces for a given point on the surface of the earth at any given moment.

To summarize: continuous changes in the position of the sun and the moon with respect to the earth cause corresponding changes in the tide-generating force. All such changes can be calculated. While the semidiurnal fluctuations due to the moon and the sun are the most important, there are a great many other less important contributory factors which cannot be neglected in the complete picture of tidal fluctuations on the surface of the earth (see Table 1).

Experimental Demonstration of the Tide-generating Forces

Newton realized that the tides were the result of gravitational forces, but in his day it was impossible to prove the presence of such forces by laboratory measurements. Only 200 years after Newton did technologists manage to perfect instruments for detecting the very small tide-generating forces of moon and sun. A freely suspended pendulum will be slightly deflected from the vertical by the horizontal component of a gravitational force, while the vertical component will increase or decrease the weight of the bob according to whether the force acts in the same or the opposite sense to the earth's attraction. However, these effects are very small indeed: the deflection from the vertical is 1/58 second of arc—a

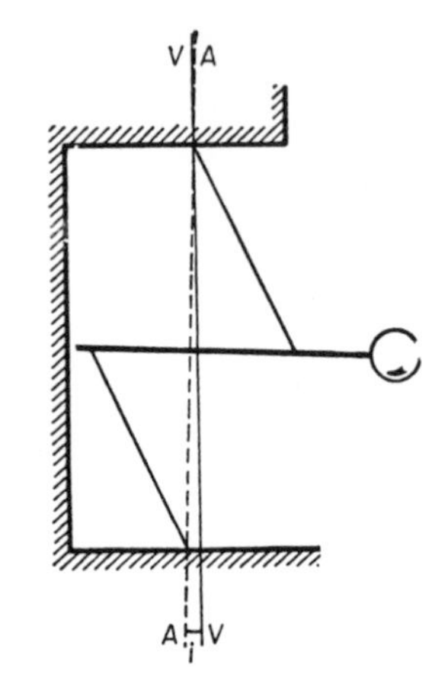

FIG. 22. The principle of the horizontal pendulum. A weighted rod is held in a horizontal position by two threads. The imaginary line *AA* makes an angle *i* with the vertical *VV*.

pendulum of 100 m would have its bob deflected by only 0.1 mm. Then again, the increase or decrease in the earth's attraction would be one millionth of its normal amount: a kilogram weight suspended from a spring and stretching the spring by one meter would be moved through an extra 1/9000 mm. Obviously such minute effects cannot be observed without suitable instruments. Such instruments do in fact exist today.

For measuring a very slight deflection, we use the horizontal pendulum. It works on the following principle: a rod weighted at one end (Fig. 22) is free to rotate about a near-vertical axis *AA* (in Fig. 22 the axis *AA* makes an angle *i* with the vertical *VV*). The pendulum will be deflected from its position of rest by any horizontal force, but once this force ceases to act the bob will automatically return to its original (lowest) position. The restoring force is $g \sin i$, where g is the earth's pull. The tractive force causing the deflection must be equal to this value. To reduce friction the axis is replaced by the bifilar suspension shown in Figure 22. The instrument is extremely sensitive to horizontal forces, but since only components at right angles to the pendulum cause it to rotate, two horizontal pendulums are needed whenever we wish to observe, for instance, the west-east and the north-south components of the tide-generating force. The rotation is so slight that it has to be magnified by an optical lever. A beam of light is directed on to a mirror affixed to the bob, and the reflected beam is recorded on a slowly moving film. For an accurate assessment of the horizontal component of the tide-generating force all other tractive forces must be eradicated. The most awkward of these are the effects of diurnal temperature fluctuations, since these effects are of the same order of magnitude as the fluctuations caused by the tide-generating forces. The instruments are therefore placed into deep mine shafts or caves where temperature changes are minimized.

Measurements of this kind were carried out first by W. Schweyder at a depth of 189 meters in Freiberg, and later by W. Schaffernicht in a cave 25 meters below ground at Marburg. Figure 23 is a graph of their results

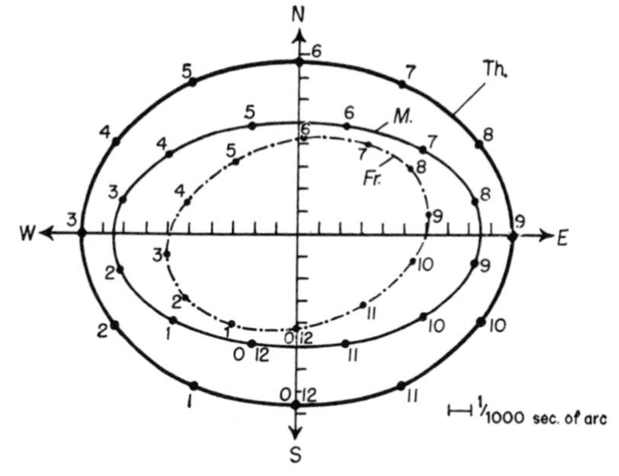

FIG. 23. Changes in direction and magnitude of the tractive force of the main lunar constituent M_2; the ellipses show the path a bob would follow during one M_2 cycle. *Th*: theoretical values; *M*: values observed in Marburg; *Fr*: values observed in Freiberg. The axes are graduated in 1/1000 seconds of arc, the ellipses in lunar hours (After Tomaschek and Schaffernicht.)

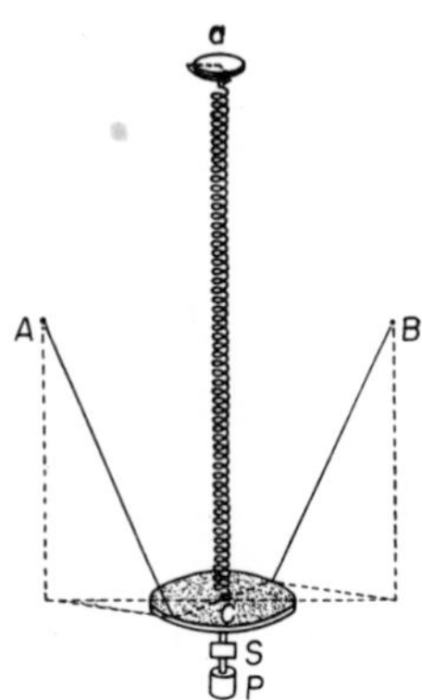

FIG. 24. Principle of the bifilar gravimeter.

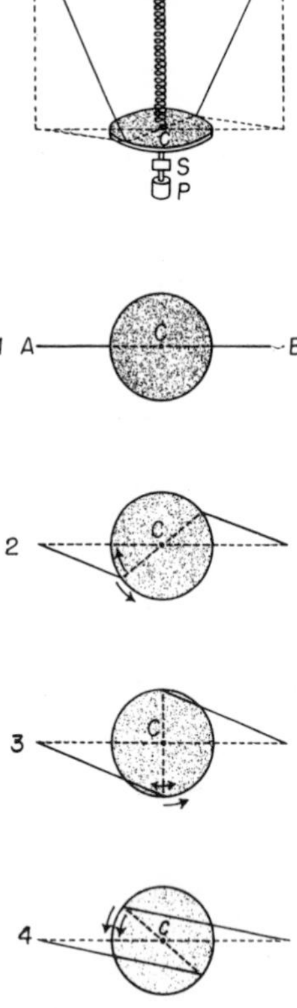

for the semidiurnal lunar tide (M_2). The deflections of two horizontal pendulums at right angles to each other are composed into resultant vectors, and a curve connecting the extreme ends of the vectors then gives the path which the bob describes under the action of the tractive force of the moon. In Figure 23 *M.* and *Fr.* are the ellipses obtained at Marburg and Freiberg, and *Th.* the theoretical ellipse. The observed elliptical paths shown in Figure 23 are smaller than the theoretical ellipse, and furthermore they show phase differences. For instance while the tractive force ought to be in the direction due north at 6^h, in Marburg this direction is reached at 5.30^h, and in Freiberg just a little before 6^h. Moreover the deflection of the pendulum is only about two-thirds the theoretical value. This discrepancy can only be due to a fluctuation with the same period as the M_2 tide. We shall return to this question in Chapter VIII.

The existence of the vertical component of the tide-generating force has also been demonstrated recently by R. Tomaschek and W. Schaffernicht using a bifilar gravimeter. Figure 24 illustrates the principle of this instrument. A disc *C* is suspended from two threads and from a spring, and has a small weight *P* attached to it. In the absence of a spring, the disc would rotate in the direction impressed upon it by the threads. Position 1 in Figure 24 shows the view from above the disc in its lowest position, the threads attached to points *A* and *B* lying in the vertical plane *ACB*. When the disc is turned by button *a*, it is pulled upward by the threads. Positions 2 and 3 show the disc being turned counterclockwise and finding after each turn a new position of equilibrium. Any further rotation of the disc (position 4) will cause the instrument to spin right round. Thus near position 3 the instrument has a very unstable equilibrium and an extremely sensitive response to small additional forces, such as the vertical component of the tidal force. This component increases or decreases the weight of the disc and

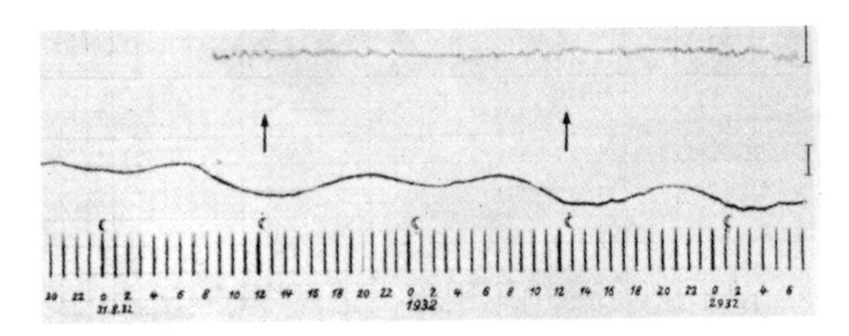

FIG. 25. Record of a bifilar gravimeter (after Tomaschek and Schaffernicht). The upper curve shows the temperature fluctuations in the given area. The sign of the moon below the lower curve tells the time of the upper or lower transit of the moon through the meridian.

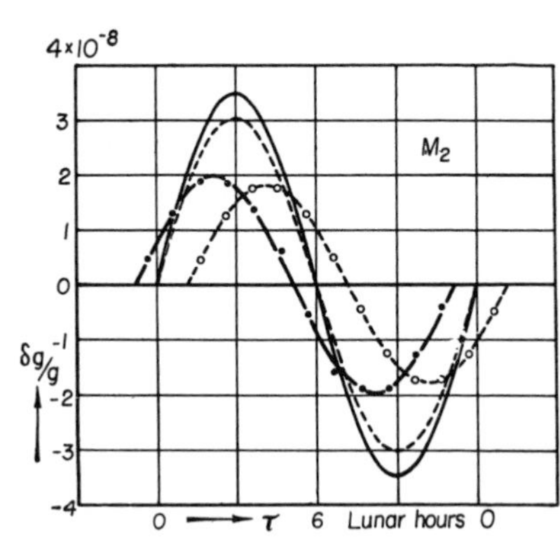

FIG. 26. Gravitational oscillations produced by the main lunar constituent M_2 in Marburg (--o---o--) and Berchtesgaden (·—·—·) compared with the theoretical values (-----) for Marburg and (———) for Berchtesgaden (bimonthly observations; moon's transit through meridian at 9^h).

of the rider *P* by a very small amount. The mirror S acts as an optical lever recording the rotation of the disc on a film. Figure 25 is such a record taken in the Marburg cave 25 meters below ground. Observations of this kind over a long time make it possible for the effects of the constituent tides to be isolated. Figure 26 is a graph of the vertical component of the main lunar tide M_2, both for Marburg and Berchtesgaden, drawn on the basis of simultaneous fortnightly observations. Once again there is a discrepancy between theoretical and actual curves, similar to that found in the case of the horizontal component of the tidal force, once again the amplitude has been reduced to two-thirds of the theoretical value and once again there are small phase differences. While Marburg lags an hour behind the moon, Berchtesgaden is one hour in advance of it. Even so, the measurements clearly demonstrate the existence of a vertical component in the tide-generating forces.

These experimental demonstrations of the tide-generating force have thus revealed a reduction in the observed force to two-thirds of their expected value. This is the result of the response of the land mass to the tide-generating force and proves the existence of a tidal motion in the earth's crust, which we will discuss in Chapter VIII.

IV. *The Origins of the Tides*

General

Tide-generating, like all gravitational, forces are proportional to the mass they attract. Now while one liter (1000 cc) of rock from the earth's crust weighs roughly 2.5 kg, the mass of one liter of sea water weighs approximately 1 kg, and one liter of air weighs only just above 1 g. Thus the tide-generating forces act on unit volume of the three substances in the ratio 2.5 : 1 : 0.001, i.e. the tides on land ought to be 2.5 times those in the ocean and 2,500 times those in the atmosphere.

However, solids are far less elastic than liquids and the very great elasticity of gases is compensated by their very low density. It is for this reason that the tide-generating forces have their greatest effect on the oceans. Even so tides occur both on land and in the atmosphere, but these tides can only be demonstrated by very sensitive instruments (see Chapters VII and VIII). Here we shall discuss ocean tides alone.

Newton's Equilibrium Theory

To obtain an approximate picture of the effect of the attractive forces on the water mass, we imagine an ideal ocean of equal depth and covering the entire earth.

Newton made this very assumption, knowing perfectly well that it was an oversimplification. To grasp the effects of the tide-generating forces on such a world ocean we must refer back to Figure 19. The horizontal component of the system of tractive forces pulls the water towards the points *Z* and *N* causing the water to heap up, with a corresponding drop in sea level along a great circle at 90° to the line joining these two points. This process cannot continue indefinitely, since the resultant horizontal pressure differences in the ocean tend to return the water to its former position. The "tidal mountains" grow until pressure differences in the ocean balance the tide-generating forces (Fig. 27). Newton's explanation of this phenomenon is known as his equilibrium theory.

Equilibrium theory considers tides as a purely statical problem. It can explain a great many characteristics of ocean tides, particularly the occurrence of semidiurnal tides. Thus, whenever the moon does *not* stand above the celestial equator, the "tidal mountains" are no longer symmetrical with respect to the earth's axis (see Fig. 20 right), and during the earth's daily rotation a diurnal inequality occurs between two tidal waves on one and the same day. However, when the moon does stand over the equator, both "tidal mountains" are equally inclined

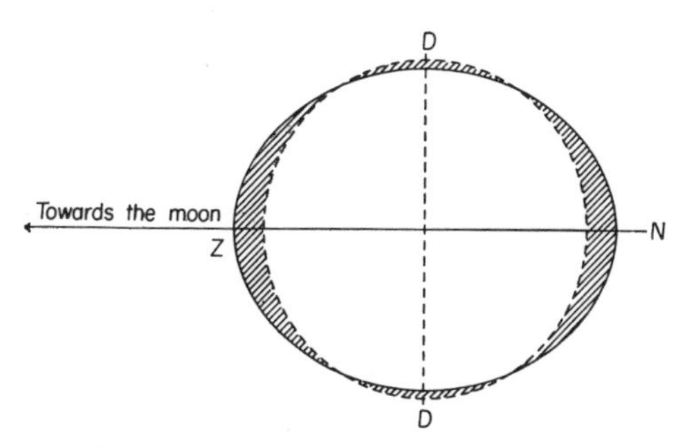

FIG. 27. Deformation of a continuous oceanic envelope caused by the tide-generating forces. The tidal mountains are fixed in the direction of the moon.

to the equator and the diurnal inequality disappears. This theory also explains how the joint effect of the tractive forces of moon and sun produces spring and neap tides, but it cannot give a satisfactory explanation for what we have called the lunitidal interval or simply the establishment. According to equilibrium theory, an observer in the Northern Hemisphere would always observe high water whenever the moon is due south above or due north below the horizon, i.e. whenever the moon passes over the observer's meridian. Thus the lunitidal interval would have to be 0 or 12 hours respectively for every place on earth, whereas observation shows the lunitidal interval to assume all values between 0 and 12 hours.

Newton knew why this was the case. The fault was not so much the assumption of an ideal ocean as the impossibility of having constant equilibrium between the tide-generating forces and the pressure. The moon changes its position with respect to the earth so rapidly that, for equilibrium to be maintained, there would have to occur enormous displacements of water within the ocean. This would have to happen at quite an impossible speed. Clearly the problem of the tides cannot be considered as a problem of statics but one of dynamics, a problem of the motion of fluids.

Laplace's Theory

A century after Newton, Laplace supplemented equilibrium theory with his dynamical theory of the tides. According to this theory the tide-generating forces, instead of producing "tidal mountains" in the ideal ocean, produce tide waves whose period corresponds to that of the forces. In the production of these tide waves other factors than the tide-generating force must also be considered: the depth and the width of the ocean basin (its morphology), the effects of the earth's rotation

(Coriolis force), and finally friction. The gyroscopic deflection due to the earth's rotation is such that all particles moving along the surface of the earth are always deflected to the right of the motion in the Northern Hemisphere, and to the left of the motion in the Southern Hemisphere. The gyroscopic effect is zero at the equator, increases with latitude toward the poles, and varies with the speed of the moving particles (e.g. the tidal current). Clearly the gyroscopic effect plays a decisive role in the formation of tidal currents and in the general picture of oceanic tides.

The water mass being shifted by the tidal currents has its speed reduced by the force of friction. In all this the usual internal friction of fluids is small compared with the friction produced by the turbulence associated with all large and wide tidal currents. The turbulent flow uses up a great deal of the energy of translation and thus has a marked effect on the final form of tides and tidal currents. We shall return to the gyroscopic and frictional effects in Chapter V. Here we shall simply give an example illustrating the influence of the depth of the water on the tidal waves. Let the moon be the only tractive force, let it move in the plane of the celestial equator, and let the water envelope consist of a narrow channel of uniform depth right along the earth's equator (40,000 km long). The two points Z and *N* in Figure 19, towards which the tidal forces are directed, would then be at a distance of 20,000 km from each other. Because of the moon's motion, these two points would move over the earth with a velocity of 1,610 km per hour, travelling right round the earth in 24 hours 50 minutes (the interval between two successive passages of the moon over a particular meridian). In such a channel the moon would then produce a tidal wave with a distance between tidal crests (wave length) of 20,000 km and with a velocity of propagation of 1,610 km per hour.

These quantities are independent of the water level in the channel.

On the other hand, a wave generated by a single disturbance of the water level (e.g. a seismic shock) would be propagated along the channel with a velocity proportional to the square root of the depth and depending on this depth alone. We find that the depth of the water in the channel would have to be 22 km for such a "free" wave to travel with a velocity of 1,610 km per hour (see page 61). Only in such a channel would the lunitidal wave keep pace with the free wave. The lunitidal force would constantly generate tidal waves all along the channel and since all such waves would travel with the velocity of the moon, the combined "free" and lunitidal waves in such a channel would reach tremendous heights. The lunitidal wave would be *in resonance* with the "free" wave. Whenever the depth of water is greater or smaller than 22 km, instead of resonance, we have interference between the two waves, causing phase differences. If the depth is greater than 22 km, high water always occurs at points Z and *N* (Fig. 19), i.e. directly below the moon, and the lunitidal interval would then be 0 to 12 lunar hours respectively as in Newton's equilibrium theory. Such tides are called "direct." On the other hand, if the depth in the channel is smaller than 22 km, troughs (low water) will occur at Z and *N*, and the lunitidal interval will be 6 or 18 lunar hours respectively. Such tides are called "indirect."

Real oceans have depths considerably smaller than 22 km, so the tides in an ocean running the length of the equator would be indirect. High water occurs 6 hours 13 minutes after the moon has crossed the meridian, and not at 0 hours postulated by equilibrium theory. Of course, this is only true for a channel running the length of the equator, but we can also calculate the lunitidal interval for other geographical latitudes. For instance, a channel at 60° would have a lunitidal wave length of ten thousand km and a velocity of propagation of 805 km per hour. Here the critical depth (depth of resonance) is 5.5 km. In the case of greater depths there would be direct tides and in the case of smaller depths indirect tides.

If we assume a world ocean of depth 10 km and if we divide it into narrow latitudinal channels, separated by walls, the channels near the equator would have indirect tides, those near the poles direct tides, while somewhere in between tremendous tides would occur because of resonance. Now if we remove the walls, the differences in water level on either side of the walls will cause currents to flow in a longitudinal direction, thus confusing the whole picture of the tides. Laplace showed that theoretically the tides would always be indirect in the equatorial regions and direct in the polar regions of such an ocean. Along a central latitude the vertical tides (but not the tidal currents) would disappear completely.

So far, we have assumed that the moon is always directly above the equator. In fact it has a declination which varies up to a maximum of 28½° north and south, and the tidal forces are not symmetrical with respect to the equator (see Fig. 20 right). The problem of wave motion in latitudinal channels thus becomes even more complicated. We shall not enter into greater details here, but it will have become clear that the dynamic theory of tides is extremely complicated. Mathematicians have computed theoretical tides for a continuous water envelope covering the entire earth, and for smaller seas bounded by lines of longitude and latitude. These theoretical solutions do not lead to a direct understanding of real tides, but they do show what factors must be taken into consideration. Nevertheless, the dynamic theory of tides has one important consequence of the utmost practical significance: it enables us to predict the course of the tides for any given place on the coast.

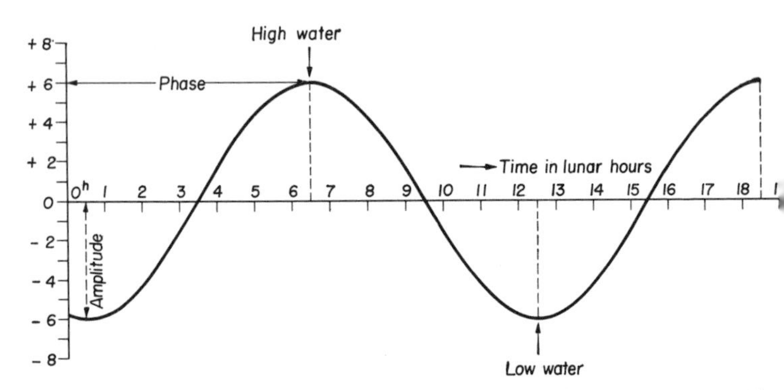

FIG. 28. Simple harmonic oscillation of the main lunar constituent M_2.
1) Time of high water: $6\frac{1}{2}^h$ lunar time
Time of subsequent high water: $18\frac{1}{2}^h$ lunar time
Period = 12 lunar hours.
2) Maximum height of crest + 6 units, Minimum depth of trough – 6 units } Range: 12 units, Amplitude: 6 units
3) Time of high water: $6\frac{1}{2}^h$ = phase.

Harmonic Analysis and the Prediction of Tides

The reader will have seen that because of (a) constant changes in the position of the moon and sun (b) constant changes in their respective distances from the earth and (c) their elliptical orbits about a common center of gravity with the earth, the tide-generating forces fluctuate periodically, i.e. they recur after regular intervals of time. Mathematicians represent such a continuous cycle as the sum of simple harmonic oscillations. In the case of tides, this is the sum of all the constituent tide-generating forces. A simple harmonic oscillation is completely determined, once we know:

1. its period, i.e. the interval between two equal phases, or in other words, the time between two maxima
2. its amplitude, i.e. half the displacement between maximum and minimum
3. its phase, i.e. the time when the maximum occurs.

Figure 28 is a graph of the simple harmonic motion of the main lunar constituent tide M_2. Its period is 12 lunar hours or 12.42 solar hours, its amplitude 6 units, and its phase 6.5 lunar or 6.73 solar hours after the moon's passage through the meridian. For an accurate picture of the tide we need to know the oscillations of all the other constituents as well, but even a relatively small number is sufficient for an understanding of the main phenomena. Each partial tide known as a tidal constituent is denoted by a letter-symbol which is an abbreviation of its main characteristic (M = moon; S = sun) and a subscript number showing whether it is diurnal (1), semidiurnal (2), or quarter-diurnal (4). There is also a host of long-period tidal constituents, but of these only the fortnightly lunar constituent (M_f) is of any great importance. Table 1 compares the periods (in solar hours) and the amplitudes of the other main tidal constituents with the main lunar constituent. According to dynamic theory the actual tide is the sum of all these tidal constituents. Because of the irregular outlines and depths of the oceans and also because of the effects of the earth's rotation and of friction, the course of the tides cannot be predicted with any accuracy, but even so, every point in the oceans, and thus every point along the coast, has a tide made up of tidal constituents, each with a period corresponding to the partial oscillation of the tide-generating force. The M_2 oscillation of the tide-generating force produces an M_2 tide with a period of 12.42 hours, the S_2 oscillation an S_2 tide with a period of twelve hours. The period of the tidal constituent can be computed from dynamic theory but not its amplitude or its phase.

TABLE 1.

The Most Important Constituents of the Tide-generating Force (Constituent Tides)

	SYMBOL	PERIOD IN SOLAR HOURS	AMPLITUDE $M_2 = 100$	DESCRIPTION
Semidiurnal tides	M_2	12.42	100.00	Main lunar (semidiurnal) constituent
	S_2	12.00	46.6	Main solar (semidiurnal) constituent
	N_2	12.66	19.1	Lunar constituent due to monthly variation in moon's distance
	K_2	11.97	12.7	Soli-lunar constituent due to changes in declination of sun and moon throughout their orbital cycle
Diurnal tides	K_1	23.93	58.4	Soli-lunar constituent
	O_1	25.82	41.5	Main lunar (diurnal) constituent
	P_1	24.07	19.3	Main solar (diurnal) constituent
Long-period tides	M_f	327.86	17.2	Moon's fortnightly constituent

However, long-term observations at a particular point on the coast enable us to isolate the amplitudes and phases of each tidal constituent from the total oscillation for the particular point. This procedure is called the harmonic analysis of tides. It must be stressed again that, while the periods of each tidal constituent are known from theory, the amplitudes and phases can only be obtained from observations. Thus for a given place, harmonic analysis leads to the determination of the tidal constants for that place.

The amplitude and the phase of a given tidal constituent do not change with time, the tides themselves having been unchanged since time immemorial. These constants are therefore characteristic for a particular coastal point, and once the relative positions of moon, sun, and earth are calculated, we can predict the tide at any future moment. The more tidal constituents are taken into consideration, the more accurate can the prediction be made. In general, and under ordinary conditions, the eight tidal constituents mentioned in Table 1 are sufficient. Harmonic analysis has led to the setting up of tide tables for a large number of international harbors. Such tables tell sailors what conditions they may expect in a given port at a given time. Figure 29 shows how great is the agreement between the computed and the actual tides. The continuous curve was computed from the seven tidal constituents (thin curves), the heavy, broken curve shows the actual course of the tide registered by a tide gauge. The agreement is excellent, but this is by no means always the case, since, apart from registering tidal fluctuations, the gauge also registers fluctuations due to wind and pressure. These two effects are particularly great during tidal surges and so are the consequent discrepancies (see Fig. 30). Figure 30*a* shows the water levels actually observed, with semidiurnal crests and many irregular fluctuations. Figure 30*b* is a harmonic analysis of the tidal

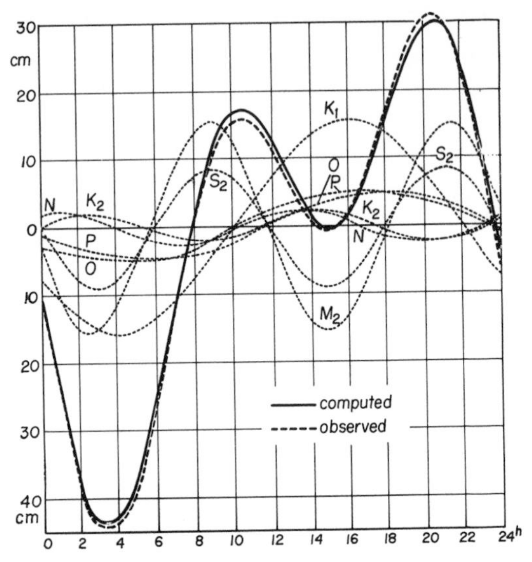

FIG. 29. Comparison between the computed and observed tides at Pula (January 6, 1909). The thin curves represent the 7 main constituent tides, the thick curve shows their resultant, and the thick broken curve the observed tide.

constituents, with the M_2 and S_2 constituents contributing most of the tidal fluctuations. In Figure 30*c* all these tidal constituents have been added and the result is a semidiurnal tide with additional diurnal and fortnightly oscillations. The difference between the observed and the computed water levels (Fig. 30*a* minus Fig. 30*c*) leaves a nonperiodic residue (Fig. 30*d*). This is the effect of wind and pressure.

The harmonic analysis of, and tide-prediction from, long-term observations is laborious and very expensive, since hundreds of coastal points must be considered year after year. This work is being done increasingly by computing machines. Figure 31 illustrates the principle on which such machines are based.

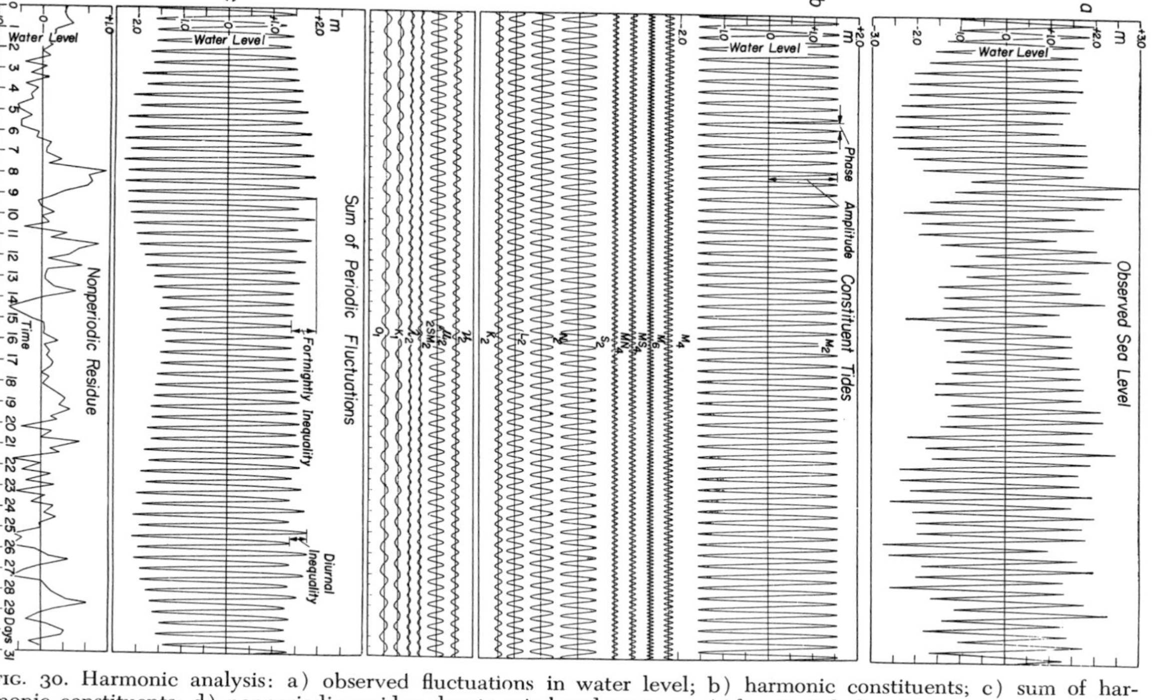

FIG. 30. Harmonic analysis: a) observed fluctuations in water level; b) harmonic constituents; c) sum of harmonic constituents; d) nonperiodic residue due to wind and pressure. (After E. Schultze.)

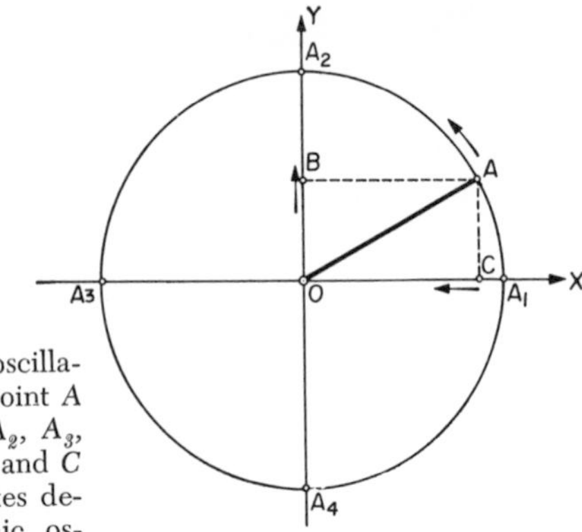

FIG. 31. Harmonic oscillation. Revolution of a point A through a circle A_1, A_2, A_3, A_4. The projections B and C on the OX and OY axes describe simple harmonic oscillations (sine waves).

Consider a point A which moves around a circle with uniform speed. The projections B and C of this point describe simple harmonic oscillations with period T along the rectangular axes OX and OY (see Fig. 28). When A is at A_1, B has zero displacement, when A is at A_2, B has reached its maximum positive displacement; when A is at A_3, B again has zero-displacement, and with A at A_4 it has reached its maximum negative displacement. Each revolution produces one simple harmonic oscillation (sine or cosine wave). The amplitude of the vibration is OA_2, i.e. the radius of the circle, and the phase depends on the time when A passes through A_1.

A machine for recording such simple harmonic oscillations would have to trace the motion of point B in Figure 31. This is done in the following way: the radius OA of the circle in Figure 31 is replaced by the crank A (Fig. 32) fitting into a slot and rotating a crossbar in a vertical plane. Each point of the crossbar including its tip B describes a sine wave during each revolution of the crank. A thread TT runs over the pulley R at B. The thread is fixed at P and holds a stylus S on the opposite side of P. As the crank moves the crossbar through a distance x the stylus S will move through twice this distance. (As the crossbar is raised by say one centimeter, the thread will be raised by one centimeter on either side of *R*, i.e. a total of 2 cm. When this happens one centimeter of thread runs over the pully from right to left.) A simple harmonic oscillation will thus be recorded on a strip of paper *pp* moving past with uniform velocity.

Now if a number of such harmonic oscillations of varying amplitude and period are required, we must have more than one such instrument. Each crank must correspond to a particular amplitude and rotate with a velocity corresponding to the period of the oscillation. For predicting the actual tide all the different individual oscillations must be added together. This, too, can be done by machine. Let R_1 in Figure 33 be one such arrangement as Figure 32; let its period be 12 hours 25 minutes, i.e. the main lunar constituent M_2. Let the thread be fastened to *P* so that during each revolution of the crank of R_1 the stylus *S* will record a sine wave. If now we make R_2 a fixed pulley and replace R_3 by another arrangement like that in Figure 32, but with a period of 12 hours (tidal constituent S_2), the stylus *S* will record a sine curve composed of two waves. There is

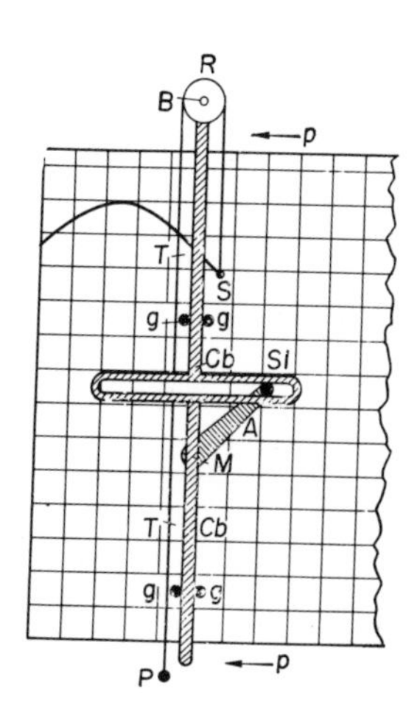

FIG. 32. Machine for producing a sine wave: *A* = Crank revolving about *M* with uniform speed, and fitting into Slot *Sl* of the crossbar *Cb*, which moves up and down between the guides *g*. The thread *TT* is fastened to *P*, runs over a pulley *R*, and carries a stylus *S*. The stylus records a sine curve on a strip of paper *p* moving in the direction of the arrows.

nothing to prevent us from using any number of such arrangements, each with an amplitude and period corresponding to the tidal constituent we wish to record. A machine for recording the resultant of all tidal constituents would have to be rather large and complicated. Its cranks would have to be adjustable both in length and in original position so as to represent different amplitudes and phases for given coastal points.

The first machine of this kind was invented by Lord Kelvin in 1872. A German machine constructed in 1919 added the oscillations of twenty tidal constituents and could draw up annual tide tables for a harbor in ten to fifteen working hours—an enormous saving of time and money. The latest tide-predicting machine of the German Hydrographic Institute in Hamburg adds the oscillations of 62 tidal constituents, although many of these are not generally needed for accurate tidal prediction. Figure 34 gives a front view of this complicated machine.

Characteristics of the Tides

Once the tidal harmonic constants for a point at the coast, i.e. the amplitudes and phases of the tidal constituents, are known, the course of the tides at that point

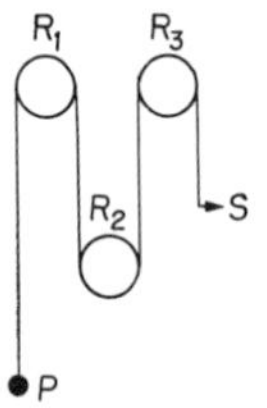

FIG. 33. The principle of the tide-predicting machine. R_2 fixed, R_1 and R_3 movable pulleys, P fixed end of thread, S stylus.

FIG. 34. The large tide-predicting machine of the German Hydrographic Institute, Hamburg. (Photograph: D.H.I.)

can be predicted. The phase of the M_2 constituents normally gives the approximate time of high water after the moon has passed through the particular meridian (the lunitidal interval or the establishment). The M_2 tide lags fifty minutes behind the S_2 tide, and the two tides will only be in phase every 14.765 days. The amplitudes M_2 and S_2 will then reinforce each other $(M_2 + S_2)$ to produce spring tides. In between two spring tides there will be one occasion when the amplitudes will oppose each other $(M_2 - S_2)$ and produce neap tides. The amplitude of the S_2 tide thus determines the fortnightly inequality. The phase difference between

S_2 and M_2 can be used for determining the "age of the tides," that is, the lag between spring tide and full and new moons.

The K_1 and O_1 constituents are diurnal oscillations associated with the moon's declination. Unlike the semidiurnal M_2 constituent they are independent of the moon's passage through the meridian. The K_1 tide gives an approximate indication (in local time) of the time of high water on June 21; each subsequent day this occurs four minutes earlier. Phase differences between the O_1 and the K_1 tides, similar to that between the M_2 and S_2 constituents, also cause spring and neap tides but these occur at intervals of 13.66 days.

The diurnal inequality, which is a consequence of the superimposition of semidiurnal and diurnal tides, varies with the ratio $(K_1 + O_1) \div (M_2 + S_2)$. This ratio, F, is an indication of the form of the tidal curve during one day. These forms are roughly as follows:

$F = 0.0 - 0.25$ (semidiurnal form). Two high and low waters of approximately the same height. High water after an almost constant lunitidal interval. Mean spring tide range is $2\ (M_2 + S_2)$.

$F = 0.25 - 1.50$ (mixed, predominantly semidiurnal, form). Two high and low waters daily, but with strong inequalities in height and phase; these reach a maximum with maximum declination of the moon. Mean spring tide range is $2\ (M_2 + S_2)$.

$F = 1.50 - 3.00$ (mixed, predominantly diurnal, form). After maximum declination of the moon, only one high water per day. Otherwise two high waters with strong inequalities in height and phase. Mean spring tide range is $2\ (K_1 + O_1)$.

$F =$ *greater than* 3.00 (diurnal form). One high water per day. Possibly two high waters at neap tide (during the passage of the moon through the plane of the equator). Mean spring tide range is $2\ (K_1 + O_1)$.

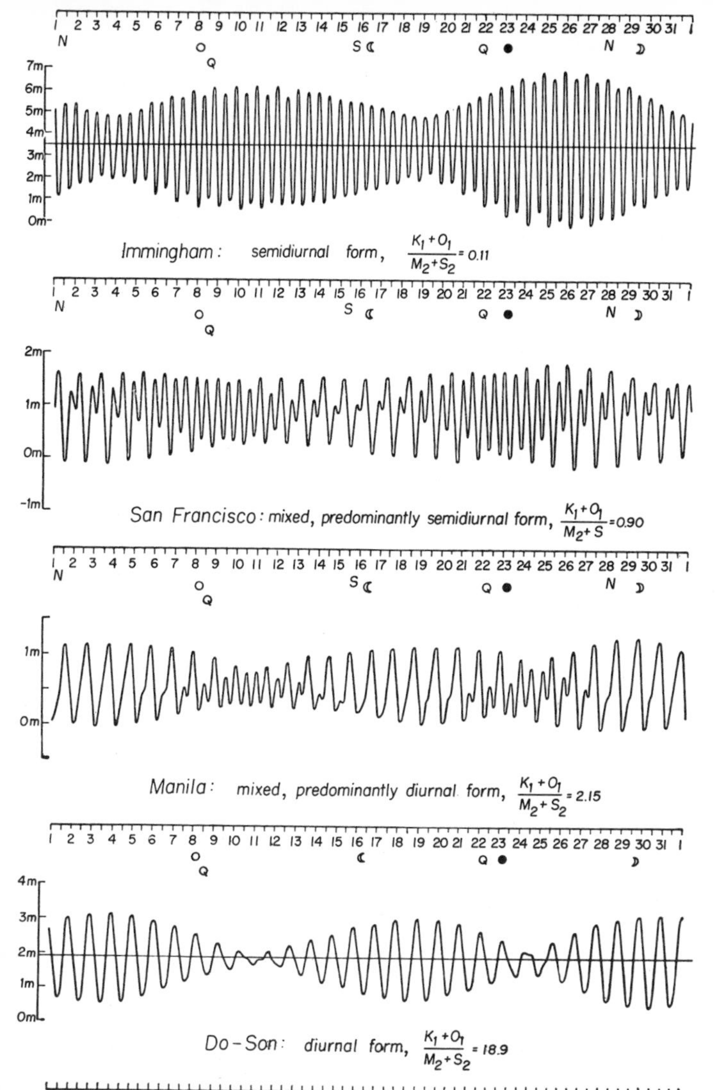

FIG. 35. Tidal curves during March, 1936 (from German Tide Tables for the year 1940, Vol. II, Berlin, 1939).
○, ☽, ●, ☾ : phases of the moon. N: maximum northern declination of the moon; S: maximum southern declination of the moon; Q: moon's transit through equator.

TABLE 2.

Harmonic Tidal Constants of the Four Harbors Whose Graphs Are Shown in Figure 35

TIDE	PERIOD	THEORETICAL AMPLITUDE	IMMINGHAM (ENGLAND)		SAN FRANCISCO (CALIFORNIA)		MANILA (PHILIPPINES)		DO-SON (VIETNAM)	
	(HOURS)	$M_2 = 100$	PHASE	AMPLITUDE IN CM	PHASE	AMPLITUDE IN CM	PHASE	AMPLITUDE IN CM	PHASE	AMPLITUDE IN CM
M_2	12.42	100.0	161°	223	330°	54	305°	20	113°	4
S_2	12.00	46.6	210°	73	334°	12	338°	7	140°	3
N_2	12.66	19.1	141°	45	303°	12	291°	4	100°	1
K_2	11.97	12.7	212°	18	328°	4	325°	2	140°	1
K_1	23.93	58.4	279°	15	106°	37	320°	30	91°	72
O_1	25.83	41.5	120°	16	89°	23	279°	28	35°	70
P_1	24.07	19.3	257°	6	104°	12	317°	9	91°	24

Figure 35 shows tidal curves for four ports during March, 1936, each having a different form number. Immingham, on the east coast of England, has a semidiurnal form with $F = 0.11$. This is the form predominating in all European harbors. San Francisco has a predominantly semidiurnal form with $F = 0.9$. Manila has a predominantly diurnal form with $F = 2.15$. Do-Son (Gulf of Tongking, Vietnam) has a typical diurnal form with $F = 18.9$.

For greater detail, Table 2 gives the harmonic constants of the most important tidal constituents for these four ports: the amplitudes are in centimeters and the phases refer to local time. The extent to which the four main constituents M_2, S_2, K_1, and O_1 govern the tides can be seen from the fact that in each of the harbors the sum of their amplitudes is roughly 70 per cent of the total amplitude. These four main constituents thus suffice for giving a fair picture of the oceanic tides.

V. *Tides in the Seas and Oceans*

Stationary Waves in Enclosed Basins; Seiches

We have already discussed the effect of the tide-generating force on a channel running along an entire line of latitude, and we saw that, while the period of the wave generated is that of the tide generating force itself, the amplitude and phase of the wave depend on the depth of the channel. Now let the latitudinal channel be of length l and depth h and let it be sealed off at both ends. The tide-generating force will then cause the water to heap up and to fall at alternate ends, i.e. the water will oscillate about a fixed central axis at right angles to the length (see Fig. 36, top). This axis is called a nodal line and oscillations about it are said to be stationary. Stationary oscillations (or waves) can have two, three, or more nodal lines in one and the same channel (cf. the overtones in a closed pipe) and Figure 36 (bottom) illustrates the water surface undergoing a binodal oscillation. While the mononodal wave always produces high and low water at alternate ends of the channel, a binodal wave produces high water at both ends and low water in the center, and vice versa.

In closed latitudinal channels the tide-generating force will always produce stationary waves with a period equal to its own, but here too amplitude and phase (time of high water) will be affected by the length and depth of the channel itself. Similarly, any *sudden* disturbance of the channel will also generate a stationary wave, since the equilibrium of the water will have been disturbed by the single short impulse. The period of this stationary oscillation of the water (produced by a single impulse) will only depend on the length and depth of the channel and is called the characteristic period of the channel. Such oscillations can be set up in natural waters (viz. in lakes, bays, etc.) by, say, a seismic shock or tidal surges, after which the water returns to horizontal equilibrium, performing successively decreasing oscillations about the equilibrium position. The period of these oscillations is the characteristic period of the channel. In a channel of length l and depth h this period is $T = \frac{2l}{\sqrt{gh}}$ where g is the acceleration due to gravity.

Now, as we have seen, a periodic force, such as one of the components of the tide-generating force, also sets up in the basin stationary oscillations whose period is that of the force and not the characteristic period of the channel. But, here too, amplitude and phase will depend on the characteristic period of the basin according to the following rules:

1. Whenever the characteristic period is very much smaller than that of the tide-generating force, there is enough time for the water level to become displaced by

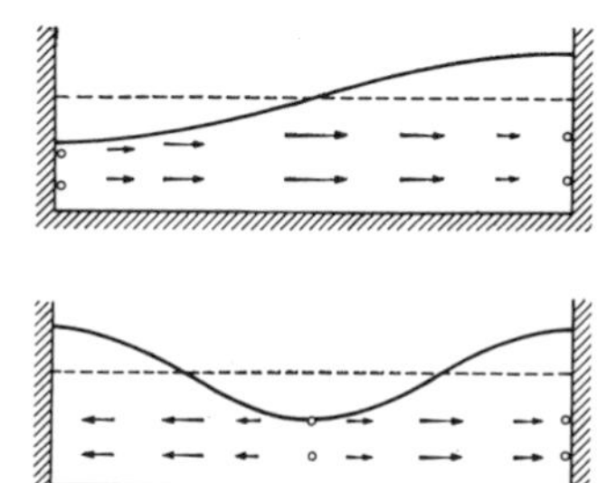

FIG. 36. Mononodal and binodal stationary waves (seiches) in a rectangular basin.

the force. The tide-generating force is thus able to impress its own period on the mass of water, and we have equilibrium tides.

2. Whenever the characteristic period is very much greater than that of the tide-generating force, the tides are small and "reversed," i.e. low water appears when the tide-generating force is maximum, and vice versa.

3. Whenever the characteristic period approximates that of the tide-generating force, the two oscillations reinforce each other. The closer the periods, the greater the tides. Such synchronous tides are called sympathetic or resonance tides.

Obviously, then, any understanding of the tides in particular seas or lakes must depend on a study of the characteristic period.

There are many factors which cause oscillations in self-contained waters; on lakes, for instance, regular, very small oscillations of the water level are quite common. Usually these oscillations can only be demonstrated by very sensitive gauges (limnimeters), but occasionally they are large enough to be noticed with the naked eye. Thus oscillations of nearly five feet have been known on the western shores of Lake Geneva in Switzerland. The local population calls these "seiches," and this word has come to refer to all such oscillations anywhere. F. A. Forel, investigating many points along the shore of Lake Geneva, has shown that seiches are in fact the characteristic oscillations of the lake as a whole. The most common such oscillations have one nodal line, but occasionally mononodal, binodal, and trinodal waves occur simultaneously with a resulting confusion of the simple picture. The causes of these oscillations are generally quick changes in air pressure, sudden squalls, etc. which, acting on a large part of the surface simultaneously, disturb its equilibrium. The moment the disturbance stops, the water level tries to return to its normal position, but oscillates about it with the characteristic period until the oscillations are finally overcome by friction. On Lake Geneva, Forel observed a series of almost 150 oscillations which, with a period of 74 minutes, lasted for a whole week.

Seiches are the characteristic oscillations of a particular lake, i.e. their period is determined by the length and depth of the lake. For lakes with a fairly regular narrow shape, the formula $T = \frac{2l}{\sqrt{gh}}$ (see p. 61) gives a good approximation for the period of mononodal seiches. Seiches with two, three, and more nodes have periods that are roughly a half, a third, etc. those of the mononodal oscillations. This formula is not applicable to any but regular lakes, although many investigators (e.g. Chrystal, Honda and his collaborators, Defant, Proudman, Hidaka et al.) have developed methods which, by means of rather laborious calculations, allow the precise computation of seiches even in very irregular lakes. The agreement between the computed and observed periods is astonishingly good. Not only the periods but even the

TABLE 3.

Comparison Between Observed and Computed Periods of the Seiches of a Number of Lakes (in Minutes)

LAKE		PERIOD OF SEICHE		
		MONONODAL	BINODAL	TRINODAL
Earn (Scotland)	obs.	14.5	8.1	6.0
	comp.	14.5	8.1	5.7
Garda (Italy)	obs.	42.9	28.6	21.8
	comp.	42.8	28.0	20.1
Geneva (Switzerland)	obs.	74.0	35.5	...
	comp.	74.4	35.1	28.0
Vättern (Sweden)	obs.	179.0	97.5	80.7
	comp.	177.9	96.0	79.2

position of the nodal lines, and also the horizontal currents associated with seiches, can be predicted. Table 3 compares observed and computed periods of seiches in a number of lakes. Clearly the agreement is excellent.

Perhaps the most careful and detailed investigation of seiches in large lakes is that carried out by Bergsten for Lake Vättern in Sweden. Its narrow uniform shape and its great length of seventy-seven miles make it particularly suited for such investigations. Figure 37 is a record of the water level at the two ends of the lake during a seiche. The seiche was mononodal and the oscillations at one end of the lake were near mirror-images of those at the other.

Now, besides the maximum vertical displacements at the crest of a stationary wave, there is also a maximum horizontal displacement of the water at the nodal line of the wave. Thus, at the moment of high water at *A* (Fig. 38*a*), and low water at *C*, there is no horizontal motion (no flow) throughout the channel. But immediately afterwards, when the water level drops at *A* to rise at *C*, there must be a flow towards the right, which is greatest at the point *B* where the gradient is greatest (Fig. 38*b*). The velocity of the flow towards the right continues to increase and is maximum when the water level is horizontal (Fig. 38*c*). Even then the flow towards the right still continues because of the inertia of the water, but it weakens as the gradient in the opposite direction builds up. At this stage (Fig. 38*d*) the water flows "uphill," so to speak, until, when the level reaches its maximum at *C*, all horizontal motion stops again (Fig. 38*e*). The whole process is now repeated from right to left. During all such oscillations:

1. The movement of the water is always and exclusively vertical at the crests, and horizontal at the nodal line.

2. There is no current whatsoever at high and low water (i.e. when the current turns), and there is maxi-

mum current when the water level passes through its equilibrium position.

Tides in Seas, Bays, and Channels

We have so far discussed an idealized model of a channel of indefinite breadth. In any real basin, however, the tide-generating force varies from place to place. Clearly, the smaller the region considered, the less this matters: the total mass of water of such small basins is then more or less subject to the same tractive force. Stationary oscillations are thus set up. Now the tidal force constantly changes its strength and direction, as illustrated in Figure 21 for the case of the main lunar constituent, M_2. At 12^h lunar time all the water tends towards south, at 3^h to the west, and so on. The crests tend to flow round a particular basin in a clockwise direction. In the case of a long, narrow channel running west-east, the tide-generating force is too weak to heap up sufficient water along its banks, and only the component along the channel (running west-east), will cause noticeable high or low waters at the west and east

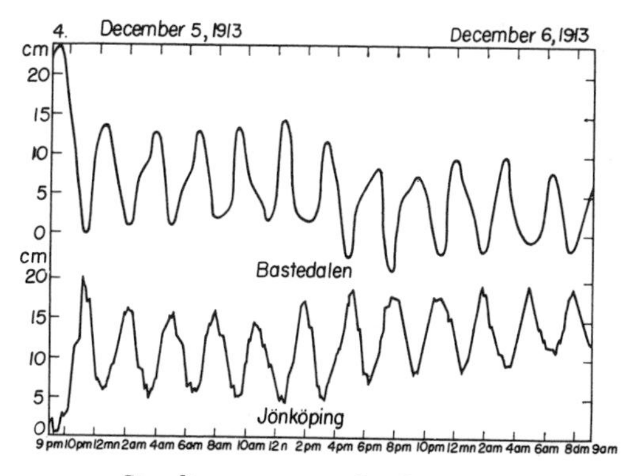

FIG. 37. Simultaneous record of a mononodal seiche in Lake Vättern (Sweden) taken at Bastedalen (northern end of lake) and at Jönköping (southern end) between 9 p.m. on December 4, 1913, and 9 a.m. on December 6, 1913.

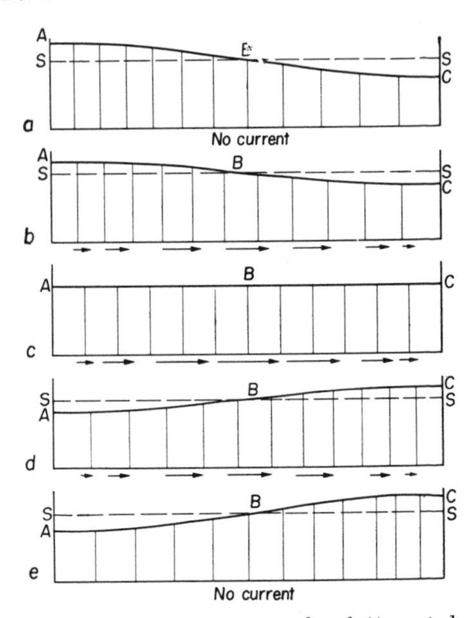

FIG. 38. Stationary waves at consecutive intervals of ⅛ period. The oscillation of the water surface and simultaneous horizontal displacements of the water are represented by the motion of the vertical columns; the volume of water between any two of these being constant. The arrows show the direction of the current and their lengths its magnitude.

ends of the channel. (This is the idealized model already discussed.) The stationary wave formed will then have a nodal line somewhere near the center. While, say, the west end will have high water all along its breadth, the east end has a phase difference of half a period, and will have low water along its breadth. Now the phase depends on the characteristic period of the basin (see p. 61). In the case of the M_2 tide (Fig. 21), a basin with a small characteristic period (small length or great depth) will have high water at the eastern end at 9^h lunar time, and at the western end at 3^h lunar time: the tides are direct. On the other hand, if the characteristic period of the basin is greater than that of the M_2 tide, the tides will

TABLE 4.

Tides in Some Lakes and Seas

LAKE OR SEA	LENGTH KM	PLACE	RANGE CM SPRING TIDE	CHARACTERISTIC PERIOD	OBSERVER
L. Geneva	61	Sécheron (western end)	0.2	74 minutes	Endrös
L. Balaton	77	Eastern and western end	5–6.5	9.4–10 hrs.	Endrös
L. Erie	378	Amherstburg (western end)	8.0	14.3 hrs.	Endrös
L. Michigan	550	Chicago	7.3	6.0 hrs.	Lenz, F. Defant
L. Superior	650	Duluth	5.9	. . .	Schureman
L. Baikal	665	Petchannaia	8.2	. . .	Sterneck
Adriatic Sea	820	Venice	37.5	21.4 hrs.	Sterneck—A. Defant
Black Sea	1191	Poti (eastern end)	8.2	5.1 hrs.	A. Defant
Baltic Sea	1475	Marienleuchte (western end)	4.0	27.3 hrs.	Witting
Eastern Mediterranean	2392	Alexandria	11.3	8.5 hrs.	Sterneck
Western Mediterranean	2020	Naples	15.0	6.0 hrs.	Sterneck

be reversed: the eastern end will have high water at 3^h and the western end at 9^h. If the characteristic period is equal to the period of the M_2 tide, resonance will occur and tremendous tides will rise up in the basin. In point of fact tides in enclosed waters, because of the small characteristic period of the basins, are generally too small to be demonstrated except by special instruments.

The smallest lake in which tides have been demonstrated is Lake Chiem in Bavaria (roughly eight miles from west to east). At its eastern shore, Endrös managed to demonstrate a range of 1 mm. Table 4 gives values for some other lakes and inland seas.

In general, seas such as the Mediterranean, the Adriatic, the North, and Baltic must be investigated in association with the oceans to which they are either connected by straits or else joined directly. From what we have said it must be clear that the form of the tidal oscillations in relatively small basins will vary with the depth and length of the basin itself.

Let us imagine a rectangular basin of even depth throughout, closed at one end and joined to the ocean at the other. Here there can be two types of tides. The tidal impulse at the junction with the free ocean will force the water in the basin to oscillate in sympathy with the external tide. These sympathetic tides or forced oscillations will be large and impressive if the basin is "synchronized" to the rhythm of the oceanic tide. The smaller the phase differences between ocean and basin, the greater are such tides. When the phase differences become very small we have the sympathetic tides mentioned earlier.

Now, the tide-generating forces also produce independent oscillations in the basin governed only by the characteristic period of the basin. Sympathetic and independent tides are the only tides found in basins open at one end. In narrow basins they always take the form of stationary oscillations with nodal lines, and lines of maximum displacement at fixed positions. The inner (closed) end is always such a position of maximum displacement. A basin closed at one end cannot sustain progressive waves, since all waves are reflected at this end. Incoming and reflected waves are superposed to produce stationary waves. Figure 39 shows cross sections of sympathetic and independent tides in basins of varying length but of constant depth. The longer the basin the more nodal lines there are. Note that the two types of tides are not identical. The position of the nodes in the two cases, and also the phases, are different. The real tides are the result of superposing the two waves, which shows how complicated are the tides in even the simplest of basins. Figure 40 shows the form of independent and sympathetic oscillations in basins of equal length but of varying depth. The top diagram is for a depth of 111 m, the second for 21 m, the third for 6 m, and so forth.

No seas are really as simple as we have assumed; they only resemble our long basins (channels) very vaguely.

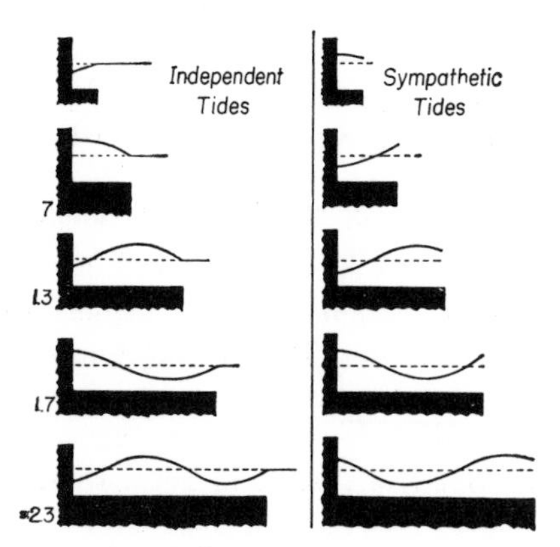

FIG. 39. Independent tides and sympathetic tides (forced oscillations) in basins of varying length but of constant depth.

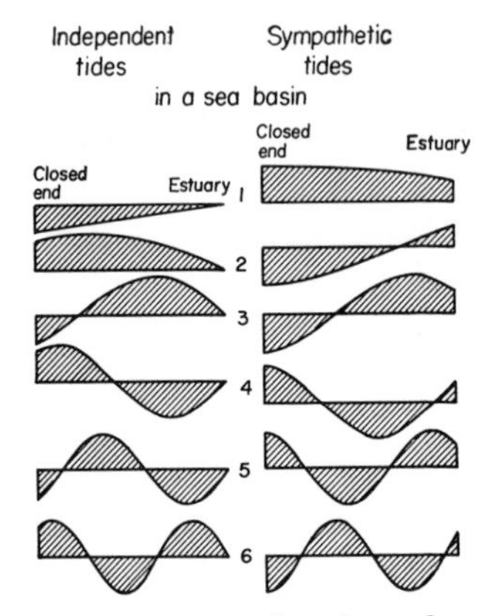

FIG. 40. Form of independent and sympathetic tides in basins of equal length but varying depth.

However, there are ways and means of calculating sympathetic and independent oscillations for even the complicated depths and outlines of the real seas. Thus the tides of a whole host of real sea basins (for instance, the Gulf of Mexico, the Mediterranean, Caribbean, Red, Adriatic, North, Irish, and Baltic seas, and the English Channel) have been analyzed and explained.

The Effects of the Earth's Rotation and of Friction

So far we have considered long, narrow basins (channels) in which we could neglect the effect of the component of the tide-generating force at right angles to the length of the basin. Now real seas usually have a breadth that cannot be ignored and hence a tidal crest moving in a clockwise direction, i.e. a rotatory tide. We shall return to this phenomenon later.

In fairly broad sea basins, there is a further factor complicating the picture of the tides. The earth's rotation produces a deflection of every motion on the earth, no matter what its direction. This force is directed to the right of the motion in the Northern, and to the left in the Southern, Hemisphere, and is known as the Coriolis force or the gyroscopic effect of the earth's rotation. It is strongest at the poles and becomes zero at the equator. The greater the speed of the motion, the greater the deflection due to the gyroscopic effect.

In the case of stationary waves in a basin in the Northern Hemisphere, the gyroscopic effect drives the water to the right or, after the tide has turned, to the left. Added to the lengthwise oscillations of both the forced and the independent tides, we now have a crosswise oscillation. It is clear that the nodal points of the lengthwise vibration will no longer form a line, but will be concentrated at a central position or area, about which the waves radiate and rotate in a counterclockwise or clockwise direction according as the basin is in the Northern or Southern Hemisphere. These no-tide points (or areas) are called amphidromic points and together with the cotidal lines radiating from them (i.e. lines having simultaneous culminations of high waters) they are called an amphidromic system.

We shall now examine the water surface and the tidal currents in an amphidromic system, by considering the elementary case of a square basin whose banks run N–S and W–E. Let it have a N–S stationary wave and a W–E nodal line through the center. Let high water at the north bank occur at 6^h (Fig. 41*a*). Let there be a superposed W–E crosswise wave due to the gyroscopic effect. The west coast will then have high water at 9^h, three hours after the north-south wave, and at a time when the north-south wave just passes through its equilibrium position (horizontal water level) (Fig. 41*d*). Thus while at 9^h all differences in water level are due to the W–E wave, they are due to the N–S wave at 6^h. The superposition of both waves which occurs, say, at 7^h and 8^h (Figs. 41*b* and *c*) causes the respective nodal lines to

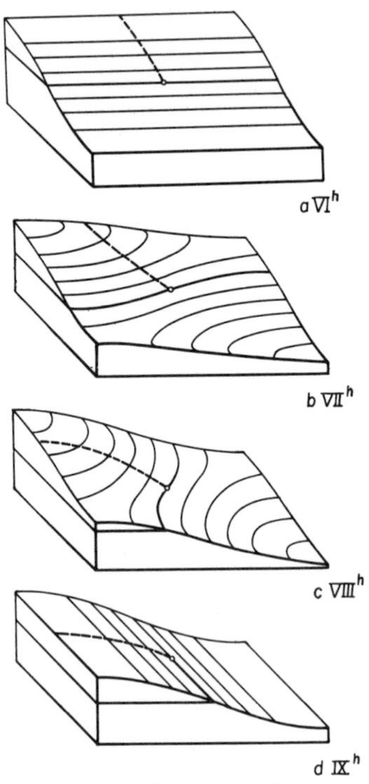

FIG. 41. Development of amphidromic system (rotatory tide) in a square basin during four consecutive hours.

disappear and to produce an amphidromic (no-tide) point at their intersection. The tide "turns" about this point, the nodal line of the E–W wave being raised and lowered by the N–S wave. Thus while the dotted line on Figure 41*a* has high water at 6^h the points west of it have not, since the E–W wave can still raise them further. Similarly, in Figure 41*d*, the dotted line alone has high water at 9^h. The cotidal lines for 7^h and 8^h (Fig. 41*b* and *c*) are intermediate. Clearly we have a rotating tide with cotidal lines radiating from a central point. Figure 42 is a map of such cotidal lines and of the distribution of the tidal range (co-range lines). The range of N–S and E–W waves respectively has been fixed arbitrarily as 100. All four banks have only one point (at the center) where the tidal range is 100, since only one of the waves affects the respective nodal line at this point. The tidal range is greatest in the corners, but since the respective waves are three hours out of phase the range at these points is 141, and not twice the central value. Maps such as these give a complete picture of the tides, since they fix the course of the tide for any given amplitude and phase. The tidal current, too, can be evaluated from the strengths of the N–S and E–W components. These are added by the law of the parallelogram of forces. At 6^h when the north bank has high water due to a crest in the N–S wave, the N–S wave itself has no internal current (see Fig. 41*a*), while the west wave, which is just passing through equilibrium position, will have a maximum E–W

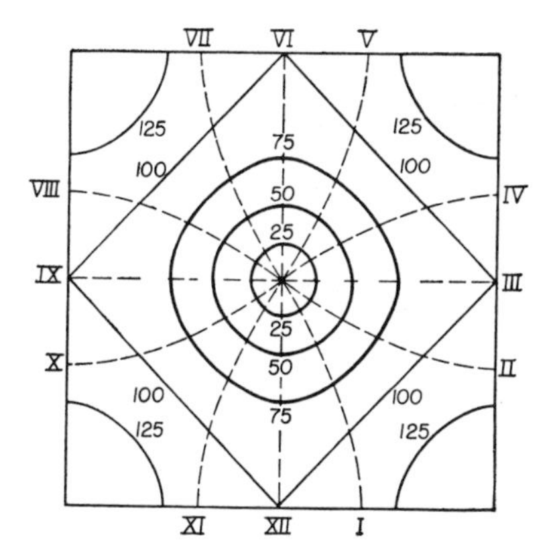

FIG. 42. Tidal map of amphidromic system shown in Fig. 41.

current (Fig. 43*a*). Three hours later, at 9^h, the whole basin will have a N–S current (Fig. 43*d*). At 7^h and 8^h the current is slowly turning (Figs. 43*b* and *c*). We have a rotatory current throughout the basin (see p. 25), and if we plot the magnitude and direction of this current during one cycle, the line joining the ends of the vectors would describe an ellipse. As we approach the banks, the ellipse becomes smaller and smaller—till finally the tidal current becomes rectilinear (alternating).

The case of the square basin is of course the simplest example, and furthermore the amplitudes of the N–S and E–W are rarely equal, as we have assumed them to be for purposes of illustration. Apart from the rotation just mentioned, there is also a rotation in the opposite direction (see p. 31). Furthermore, in irregular basins the two component waves are usually of unequal strength. The cotidal and co-range lines will no longer be as regular as in Figure 42, but even so the cotidal line will

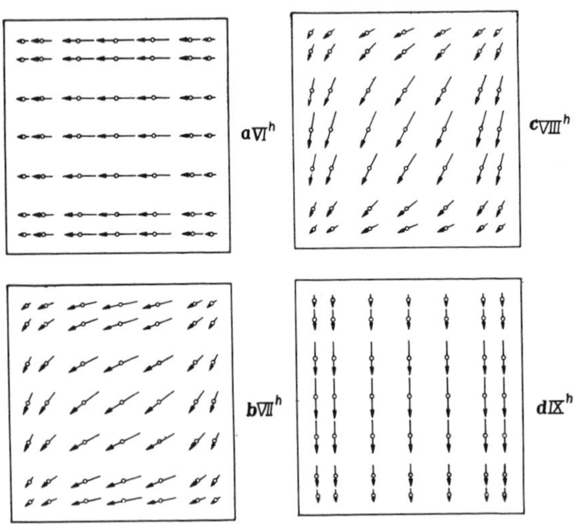

FIG. 43. Hourly changes in current of the amphidromic system shown in Fig. 41.

always pass through an amphidromic point and the co-range lines will always form closed curves about this point.

Yet another factor can influence the tides: friction. While friction is unimportant compared with the gyroscopic effect, it still results in a reduction of the tidal range and also in a phase lag in the time of high water. This friction is due to the turbulent flow of the tidal currents when there is an uneven sea bed in a shallow sea or when the coast is very irregular. The current then gives rise to eddies and rollers which use up a considerable part of the tidal energy. The shallower the waters, the greater the effects of such friction. In the case of stationary tidal waves, so characteristic of more or less enclosed sea basins, the visible effect of friction is confined to near the nodal lines. The greater the friction, the more does a particular nodal line turn into a type of progressive wave. The change in phase is most noticeable at the original position of the nodal line. There is a simultaneous reduction in the tidal range with distance from the open ocean, the reduction in amplitude being maximal at the inner end of the basin.

Tides in Specific Waters

The Red Sea, a deep and narrow basin, and the Gulf of Suez, a shallow and narrow basin.—The Red Sea so resembles a long, narrow, and straight channel that there can be no crosswise oscillations of any great importance. Its great depth (average 1,560 feet) assures minimum friction. All the theoretical assumptions seem to be satisfied. At its southern end the Red Sea joins the Gulf of Aden and the Indian Ocean by the narrow Strait of Bab el Mandeb. In the north, the Red Sea runs into two different types of narrow channels (Gulf of Suez and Gulf of Aqaba). The former is shallow (average depth 118 feet), the latter deep (average depth 2,135 feet). Both must therefore have quite different tides. Let us first look at the Red Sea itself. Its comparatively small outlets to the north will not greatly affect its own tides, which will therefore be a superposition of its characteristic, independent tides and of the sympathetic tide from the Gulf of Aden (Indian Ocean). Both can be calculated very accurately from the shape of the basin. The sympathetic tide will be roughly the fourth type shown in Figure 40 (right), i.e. with two nodes, one near the mouth and one in the center of the basin. At the estuary of all basins (and here at the junction with the Gulf of Aden) there is always a nodal line (see Figs. 39 and 40), for at such junctions the water can enter and leave freely, a process which favors the formation of nodal lines. The independent tide in the Red Sea will be the fourth kind shown in Figure 40 (left): a node at the mouth and a second in the center of the channel. Figure 44 shows the two component tides for the main lunar constituent M_2. Clearly theory and observation are in excellent agreement. At the mouth, the sympathetic tide is roughly three times as great as the independent tide. The actual range is not the result of simply adding the component tides, for, while the high water of the independent tide

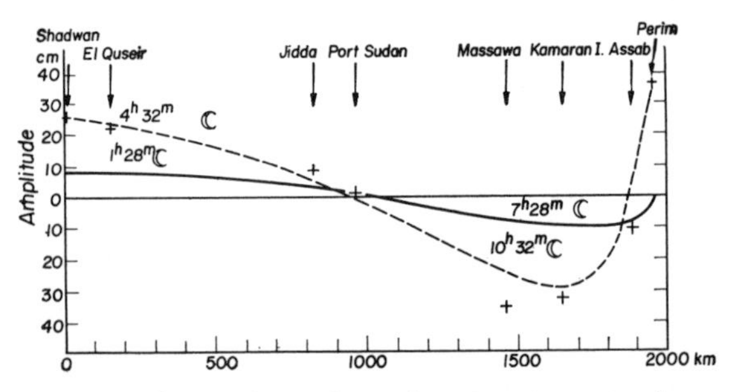

FIG. 44. Tides in the Red Sea between Perim in the south and the Gulf of Suez (Shadwan) in the north. The crosses show the observed amplitudes, the continuous curve the independent tide (caused directly by the attraction of the moon), and the broken curve the sympathetic tide with the Indian Ocean. The times of the respective phases are in lunar hours. (After Grace.)

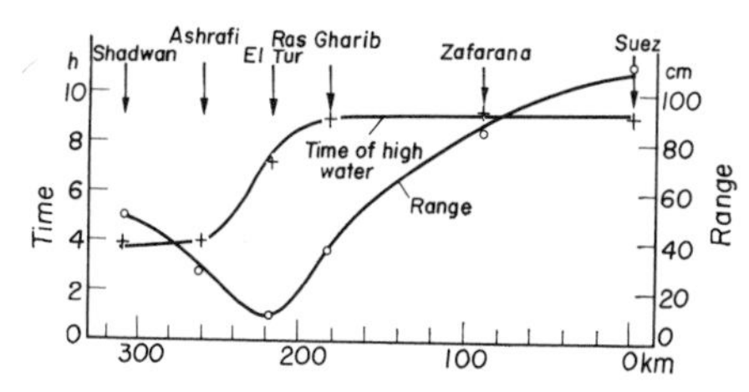

FIG. 45. Time of high water (-+-+-+-) and tidal range (-o-o-o-) in the Gulf of Suez; times refer to transit of moon through Greenwich meridian. (After Grace.)

is determined by the passage of the moon (or the sun) through the meridian, the sympathetic tide is determined by high water at the estuary. The maxima of the two partial tides are never synchronous—they damp each other. Figure 44 shows a phase difference of roughly three hours.

While the Gulf of Aden has marked diurnal tides, the Red Sea is not affected by these. The Red Sea has no resonance with 24-hour waves, so that the amplitude of the diurnal tides remains rather small. Only near the nodal lines of its usual semidiurnal tides (Port Sudan and Mocha) are any diurnal effects observable, since here the semidiurnal tides have become reduced to very small values.

The Gulf of Suez has almost the same shape as the Red Sea but is only about a sixth of its length. Its average depth is $1/11$ of that of the Red Sea, so that friction will have a marked influence on its tides. Here, too, the main component tides will be (a) the sympathetic tide with the tide of the Red Sea at the junction (near Shadwan) and (b) the independent tide. However, because of the small depth and length of the Gulf, the independent tide will be far less important than the resonance tide. Observations (Fig. 45) show that, from the junction near Shadwan, the range first diminishes to reach its minimum value near El Tur. Then there is a rather quick and strong increase in the range as far as Suez at the inner end. If the tidal range near El Tur were zero and if the time of high water were to jump by six hours, the tide would have the character of a stationary wave. But, as Figure 45 shows, this is only partly the case. These deviations from a purely stationary wave must be due to friction. Normally, at resonance, the progressive wave would enter through the junction and, in the absence of friction, would run the length of the channel to the closed end where, without change of amplitude, it would be reflected back to the junction. The incident and reflected waves would become superposed, and the result would be a stationary wave. If, however, there is friction, the incoming wave will become damped on its way in, only partially reflected, and further weakened on its way out. The resultant wave can never be a purely stationary wave. The original nodal line of the sympathetic wave (forced oscillation) is no longer a no-tide line, while at this point the phase undergoes a rapid change. This is precisely what happens near El Tur, so that the tides of this shallow narrow basin are unquestionably sympathetic tides with the tides of the Red Sea, but sympathetic tides modified by friction.

The other arm of the Red Sea, the Gulf of Aqaba, is almost the same length as the Gulf of Suez, but on the average almost 150 times as deep. Thus its characteristic period is very small and its tides will be the first sympathetic type of Figure 40. Its waters will simply oscillate in sympathy with those of the Red Sea at its mouth. Observations confirm these theoretical conclusions. It is striking how two basins in such close proximity and superficially so similar have such different tidal behavior because of different depths.

The Adriatic, a deep, broad basin.—The Adriatic cuts deep into the European mainland. It has the shape of a long basin of almost constant width and joins the Ionian Sea through the Strait of Otranto. An oceanographic map of the Adriatic shows two basins of uneven depth and form. The northern basin has a maximum depth of 790 feet near the barrier separating it from the southern basin and flattens out towards the Gulf of Trieste in the north. The southern basin is like a funnel whose deepest point is 4,100 feet. A second barrier in the Strait of Otranto separates this funnel from the Ionian Sea. The tides in the Adriatic are very well known, since harmonic constants have been established for a large number of coastal points. The semidiurnal tides are characterized by a well-developed amphidromic system in the northern part, while the diurnal tides are completely synchronous with the tides of the Ionian Sea, and have an increase in amplitude towards the north. The oceanographic contours of the Adriatic would lead one to expect that the sympathetic tide would be a lengthwise oscillation with a nodal line passing near Zadar. The outer parts of the Adriatic would then have high water when the inner ones have low water and vice versa. The seimdiurnal tides (e.g. M_2 or S_2) do in fact behave in this way, but the existence of a counterclockwise rotatory tide clearly shows that there are crosswise oscillations with a phase difference of three hours from the simple lengthwise wave. As was seen on page 70 such crosswise oscillations are due to the gyroscopic effect of the earth's rotation. Now, every simple wave can be considered as the resultant of two waves whose phase (time of high water) differs by a quarter period (3 lunar hours in the case of the M_2 tide-period 12 lunar hours). At every coastal point, the observed tide can thus be resolved into two components: one lengthwise and the other crosswise. Both can be calculated theoretically once depth and breadth are known, which is the case in the Adriatic, and tidal theory can be put to a very good test in these waters. Figure 46 (top) shows the theoretically computed amplitude of the lengthwise oscillations of the main lunar tide M_2. Both the very pronounced sympathetic tide and the very weak independent tide were taken into consideration. The opening of the Adriatic has high water at 3.3^h lunar time, the innermost part at 9.3^h lunar time. Thus the sympathetic tide is a simple oscillation about a nodal line some 290 kilometers (180 miles) from the northern end. The crosses on the graph mark the observed components of this lengthwise oscillation and we can see that they are in very close agreement with theory. The tidal currents associated with the lengthwise (N–S) component are deflected due to the gyroscopic effect. The resulting crosswise oscillations also can be roughly predicted by

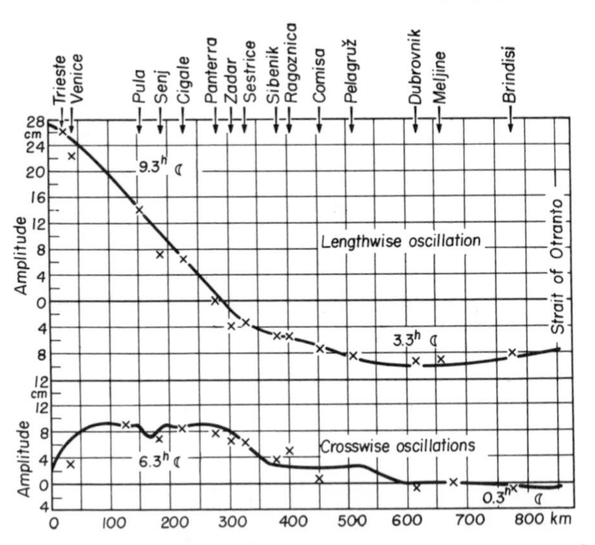

FIG. 46. Tides in the Adriatic; main lunar tide M_2 (period 12 lunar hours): the upper curve shows the theoretical forced oscillations due to the impulses from the Ionian Sea; the lower curve shows the theoretical crosswise oscillations as the result of the gyroscopic effect on the lengthwise oscillation. The crosses are the lengthwise and crosswise components of the observed oscillations at all the coastal points mentioned.

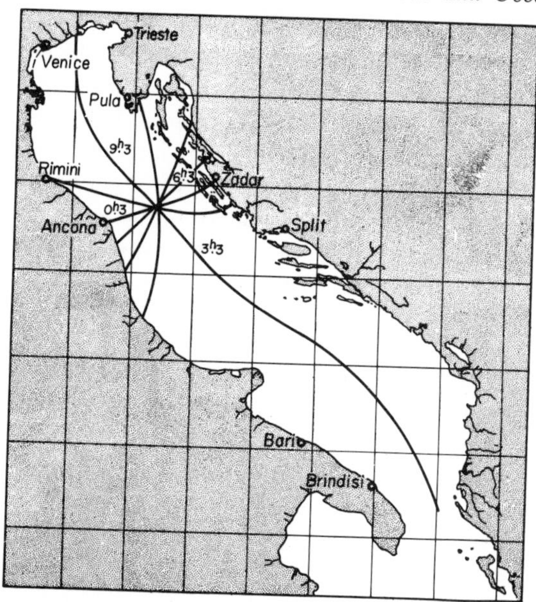

FIG. 47. Cotidal lines of the Adriatic M_2 tide. The times are in lunar hours. The amphidromic system is the result of the gyroscopic effect of the earth's rotation (Coriolis force).

theory once the strength of the lengthwise tidal current is known. Now these currents can be determined from the amplitude of the lengthwise oscillation. In the lower part of Figure 46, these computed values also are compared with the observed values for the M_2 tide. Here, too, the agreement is excellent, particularly if we remember that every point is influenced by special local factors. Lengthwise and crosswise oscillations together lead to the total tidal picture of the M_2 tide shown in Figure 47. Similar tidal maps can be drawn for the other semidiurnal tide constituents (e.g. S_2), all of which are characterized by a well-formed amphidromic system.

The diurnal tides, e.g. the K_1 tide, can also be calculated on the same theoretical basis, and here too there is excellent agreement with observations. The diurnal tide is synchronous with the corresponding tide in the Ionian Sea. It has no nodal line and thus no rotatory tide. Furthermore these diurnal tides are very weak, and hence completely obscured by the gyroscopic effect. In short, we may say that the tides in the Adriatic are not so much independent as due to semidiurnal periodic impulses from the Ionian Sea, their character depending on sympathetic oscillations modified by the gyroscopic effect.

The North Sea, a broad basin.—The North Sea differs from the Adriatic in that it is broad compared with its length and that it has a very wide junction with the open ocean in the north. Also, the tidal range at many points along its coast is far greater than in the case of the Adriatic. On the English coast, for instance, there is a mean range of between 9.5 and 16 feet. On the other hand, the range falls to the very low value of 10 inches and less along the Norwegian coast.

Because of the breadth of the North Sea, the theoretical approach we have used in the case of the Adriatic can no longer be applied directly. But, as a first approximation, the North Sea can be considered as a broad basin open to the ocean along its entire northern side, thus receiving strong impulses from the external tide of the Atlantic Ocean. The southwest opening, i.e. the Straits of Dover, will have no great influence; because of the small volume of water passing through these Straits they can only influence the tidal picture in a narrow part between England and Holland. In the North Sea the only significant tides are semidiurnal, diurnal tides being so small that they can be ignored for ordinary purposes. The main lunar constituent M_2 can be taken as representative of all semidiurnal tides. Figure 48 is a tidal map for this constituent. It shows how the tides enter the North Sea from the north as a progressive wave. Its velocity of propagation is considerably greater along the Scottish and English coasts. Here there are also considerably greater tidal ranges than on the Norwegian coast, where the progress of the wave is clearly impeded. Just off the Norwegian coast near Stavanger and Kristiansand there is a small amphidromic point. Its existence has long been questioned, but the most recent investigations seem to establish it beyond doubt. At this point the amplitude of the tidal wave is minimal. The wave travels south along the western side with great amplitude, then turns

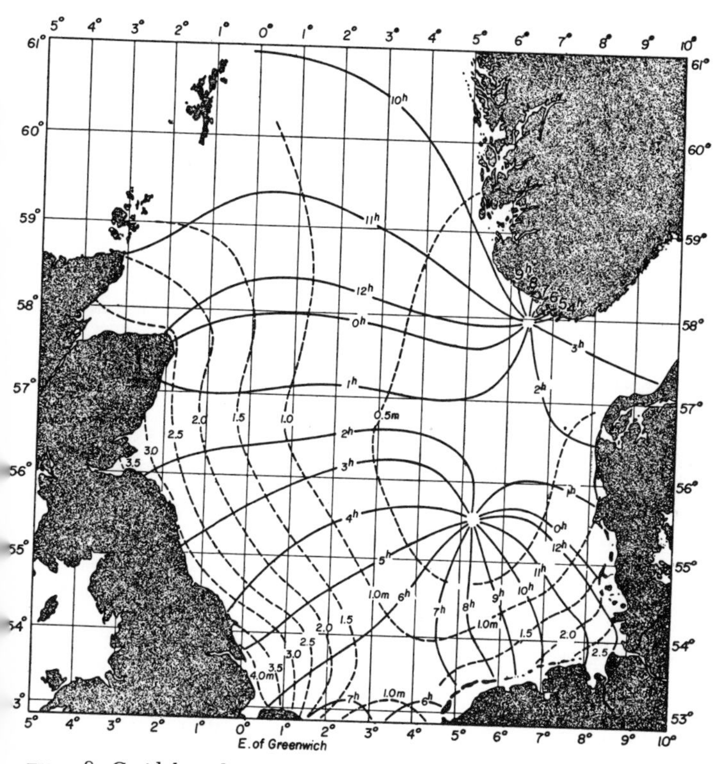

FIG. 48. Cotidal and co-range lines of the North Sea M_2 tide (after W. Hansen). ———: time of high water after moon's transit through Greenwich meridian; ---------: mean co-range lines.

north along the eastern side, finally closing about the amphidromic point off Jutland.

Now, tidal observations of the North Sea are rather sparse and hardly sufficient for an accurate tidal map of the entire region. However the tidal currents in the North Sea have been studied at length. In two ways it is possible to derive a tidal map from the strength of such currents. In the first method a large region of the sea is divided into a number of smaller areas, for each of which current observations show the difference between incoming and outgoing volumes of water per unit time. From this the range and the tidal curve can be derived. In the second method theoretical relations between drop in water level and current velocity are used to derive the tidal curve at sea from data obtained by observations at the coast. For the North Sea, both methods have led to concordant results, and it is on these that Figure 48 is based.

The North Sea tides are in all probability mainly sympathetic with the Atlantic tides at the junction in the north. The length and depth of the North Sea would lead one to expect two nodal lines, one in the north between Scotland and Stavanger, the other across Heligoland Bay in the south. The fact that the amphidromic points lie just in these regions would make this simple explanation a very probable one. The tides' rotation to the left is probably due to the gyroscopic effect. However, while in the case of the rather narrow Adriatic the crosswise oscillations due to the gyroscopic effect could be simply derived from the theoretical flow of the lengthwise wave, this cannot be done in the case of the very broad North Sea. Closer investigation of this problem leads to the following question: what happens when a progressive tidal wave of given phase and amplitude moves through a broad channel and is reflected from the closed end? For an answer we must investigate the behavior of a progressive tidal wave in a broad channel under the influence of the gyroscopic effect. In the crest of such waves the flow is in the direction of propagation of the wave, but in the trough it has the opposite direction. The gyroscopic effect in the Northern Hemisphere will thus turn the flow in the crest to the right of the direction of propagation, and the flow in the trough to the left. The tidal ranges will differ on either side of the channel (see Fig. 49). In the absence of the gyroscopic effect, let the progressive wave in a broad channel have high water H and low water h; then $H - h =$ the tidal range. Now the gyroscopic effect will cause the water level during high tide to be deflected to the right side until the gyroscopic effect is balanced by the resulting pressure differences in the water. While the high water level on the right-hand side mounts to $H + a$, it drops to $H - a$ on the left. During low water the gyroscopic effect acts in the opposite way: the water level during low tide will be raised to $h + a$ on the left side, while it will be lowered to $h - a$ on the right side. The range on the right-hand side is therefore $H - h + 2a$. The gyroscopic effect has produced an increase of $2a$ on the right and a decrease of $2a$ on the left so that there is a change of $2a$ in the range on either side. Figure 50 illustrates this type of progressive wave, called a Kelvin wave. Here all motion across the channel has ceased, since there is constant equilibrium between the gyroscopic force

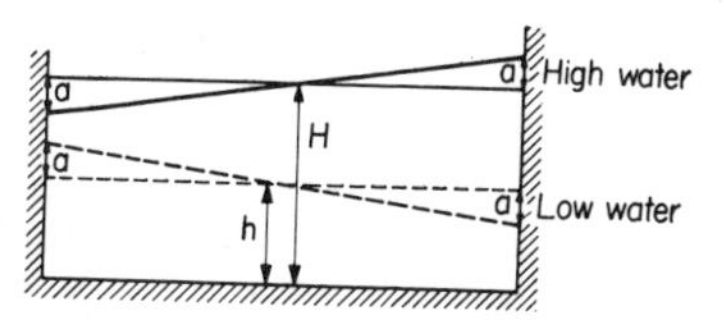

FIG. 49. Effect of the Coriolis force on a progressive tidal wave in a channel. Direction of propagation perpendicular to plane of paper. Horizontal lines: water level during high and low water in the absence of a gyroscopic effect. Skew lines: displacement of water level due to gyroscopic effect of the rotation of the earth (Coriolis force).

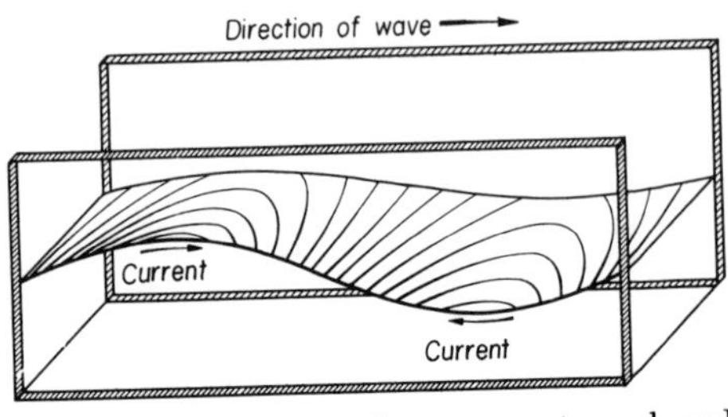

FIG. 50. Progressive Kelvin wave in a broad channel under the influence of the Coriolis force (Northern Hemisphere). The lines on the water surface are water level contours; they show the surface of the sea during the passage of the wave.

and the pressure gradient. Broad channels on a rotating earth can only have Kelvin waves.

Now whenever such Kelvin waves reach the closed end of a channel, they are reflected completely. However, this process is no longer so simple as in the case of an ordinary tidal wave. G. I. Taylor has shown that in this case there appear special crosswise oscillations of the water which completely alter the tidal picture. Only at some distance from the innermost (closed) end is the tide the simple superposition of the incoming and reflected Kelvin waves. For all channels with dimensions such as the North Sea, the tidal picture must roughly be that shown in Figure 51. There must be two rotatory tides, one in the outer and the other in the inner part of the channel. The tidal wave must run from north to south along the western side (if the open end of the channel is in the north), rotate along the closed south and then return to the north by way of the eastern side. If we compare these theoretical considerations with the actual tidal picture of the North Sea, we are struck by a great similarity. Still, there are some discrepancies that require explanation. Why are both rotatory tides deflected towards the east, the external more so than the

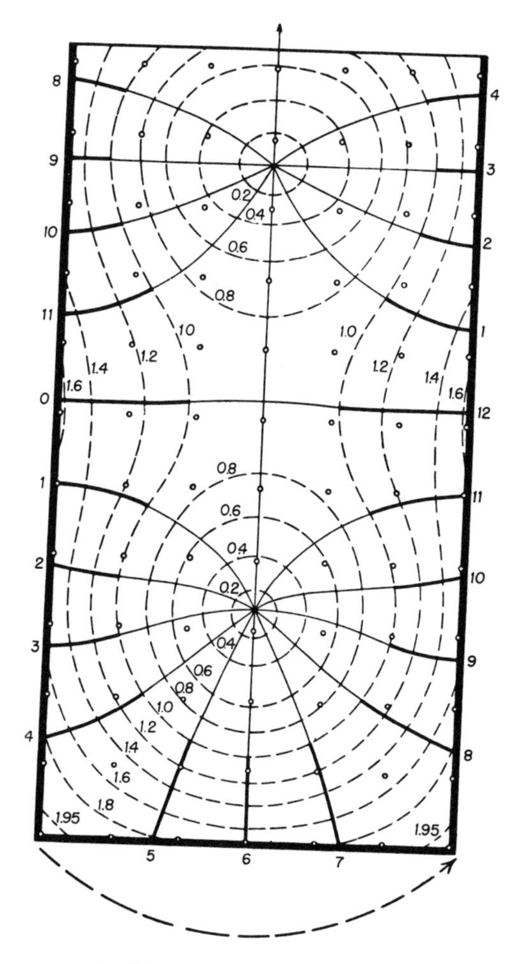

FIG. 51. Cotidal lines and co-range lines of the semidiurnal tide in a bay whose length is twice its breadth (after G. I. Taylor). The arrow shows the direction of the earth's rotation.

internal? Why is the internal far more highly developed than the external? The answer is friction. The North Sea being very shallow, the tidal streams are affected by the unevenness of the sea bed. The incoming Kelvin wave is weakened even on its way in, but the greatest loss in tidal energy certainly occurs at the innermost end of the North Sea, which is shallow and has off-shore islands and extensive mud flats. The reflected Kelvin wave is from the very start much weaker than the incoming wave and is weakened further on its way out. The superposition of the two waves is just big enough to produce a rotatory tide at the interior of Heligoland Bay, but even so its center has already been displaced slightly towards the east. In the case of the external amphidromic system, the reflected wave is so weak as to produce only a small rotatory tide in the east. Thus along the entire breadth of the North Sea, the tidal picture is almost exclusively that of the incoming Kelvin wave.

At root, then, the tides in the North Sea are determined by the Atlantic Ocean. The North Sea receives a tidal impulse at its northern opening and sets up sympathetic tides. While the gyroscopic effect and friction have a decisive influence on the formation of this tidal picture, independent tides and impulses from the Strait of Dover are insignificant, and so is the loss of energy through the Skagerrak.

Tides in the Open Ocean

We should be able to understand the tidal processes on the open oceans if only we knew the phase and amplitude of the tidal wave for enough points of the ocean. Unfortunately this is not the case. Observation has led to the determination of tidal constants at many coastal points and islands, but it is by no means clear how these constants affect the wide tracts of the open ocean itself. A whole host of interpretations can be offered, and it is important to investigate which of these

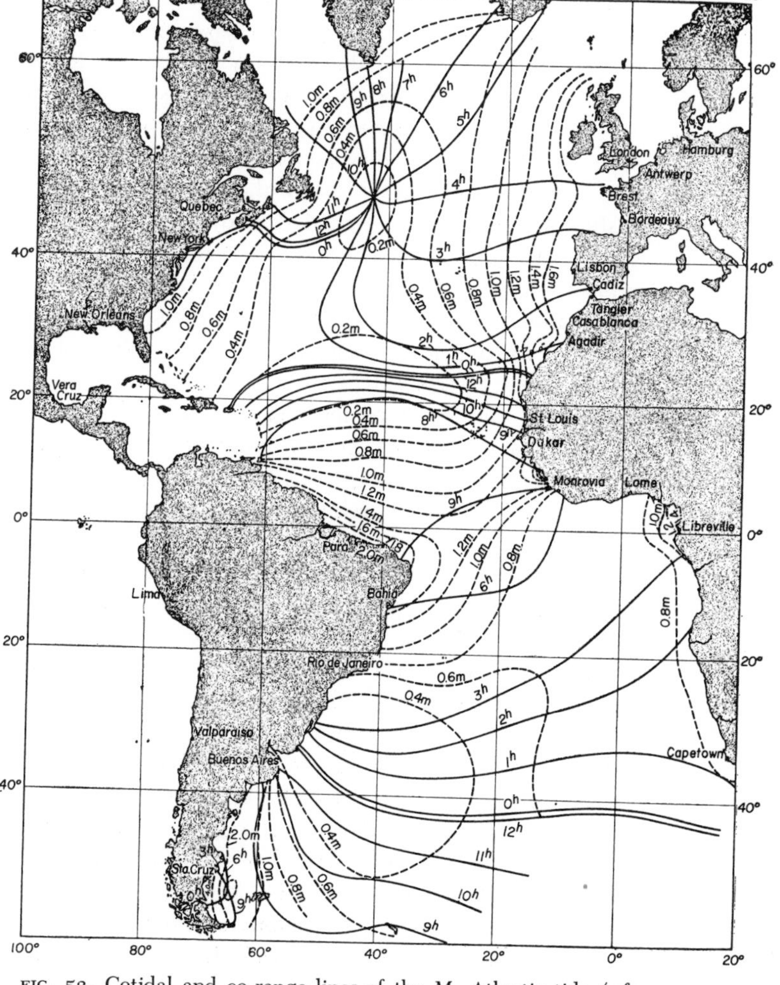

FIG. 52. Cotidal and co-range lines of the M_2 Atlantic tide (after W. Hansen). The times of high water refer to the moon's transit through the Greenwich meridian, the co-range lines are in meters.

are reasonable and have some degree of probability. Many such attempts have been made (Whewell, 1836; Harris, 1904; von Sterneck, 1920). The latest, made by Dietrich in 1944, is based on a critical evaluation of all the available observational material. Even so, his picture is more or less conjectural, and the maps are unlikely to present more than a probable state of oscillation for the types of tides considered. Figure 52 shows the cotidal lines of the M_2 constituent tide in the Atlantic Ocean (after W. Hansen). The behavior of this constituent is, of course, more accurately known than that of any other constituent. The times are related to the moon's passage through the Greenwich meridian. The amplitude distribution has been calculated theoretically from values obtained at coastal points, and these are obviously affected by friction arising from coastal and sea bed contours (see p. 75). Table 5 gives tidal constants for the main semidiurnal M_2 and S_2 tides and for the most important diurnal K_1 and O_1 tides, at a number of points on the west and east coasts. The phases (in degrees, $360° = 12$ lunar hours) are again related to the moon's passage through the Greenwich meridian.

Along the entire width of the South Atlantic, a tidal wave advances from south to north (see Fig. 52). Cotidal lines fan out from the South American to the South African coast and off Rio Grande do Sul there is a compression of cotidal lines together with an amplitude minimum (5 cm). A jump in phase of six hours on either side of this compression makes it probable that this is a noncharacteristic nodal line. In the equatorial region there are small phase differences, large amplitudes, and a further compression of cotidal lines crossing the ocean at roughly 20° N. Here, too, the amplitudes are small, so that this compression is probably another nodal line. The entire North Atlantic has a well-developed counterclockwise amphidromic system due, no doubt, to the gyroscopic effect. At the junction with the North Sea

TABLE 5.

Some Tidal Constants in the Atlantic

PLACE	LAT.	LONG.	AMPLITUDE IN CM				PHASE IN DEGREES				FORM NUMBER
			M_2	S_2	K_1	O	M_2	S_2	K_1	O	%
St. John's (Newfoundland)	47°34′ N	52°41′ W	35.7	14.6	7.6	7.0	210	254	108	77	29
New York (Sandy Hook)	40°28′	74°01′	65.4	13.8	9.7	5.2	218	245	101	99	19
St. George (Bermuda)	32°22′	64°42′	35.5	8.2	6.4	5.2	231	257	124	128	27
Port of Spain (Trinidad)	10°39′	61°31′	25.2	8.0	8.8	6.7	119	139	187	178	47
Pernambuco	8°04′ S	34°53′	76.3	27.8	3.1	5.1	125	148	64	142	8
Rio de Janeiro	22°54′	43°10′	32.6	17.2	6.4	11.1	87	97	148	87	35
Buenos Aires	34°36′	58°22′	30.5	5.2	9.2	15.4	168	248	14	202	70
Moltke Harbor (S. Georgia)	54°31′	36°0′	22.6	11.7	5.2	10.2	213	236	52	18	45
Capetown	33°54′ S	15°25′	48.6	20.5	5.4	1.6	45	88	127	243	10
Freetown	8°30′ N	13°14′	97.7	32.5	9.8	2.5	201	234	334	249	9
Puerto Lux (Las Palmas)	28°9′	25°25′	76.0	28.0	7.0	5.0	356	19	21	264	12
Ponta Delgada (Azores)	37°44′	25°40′	49.1	17.9	4.4	2.5	12	32	41	292	10
Lisbon	33°42′	9°8′	118.3	40.9	7.4	6.5	60	88	51	310	9
Brest	48°23′	4°29′	296.1	75.3	6.3	6.8	99	139	69	324	5
Londonderry	55°0′ N	7°19′ W	78.6	30.1	8.2	7.8	218	244	181	38	15

a nodal line reaches out into Denmark Strait (between Greenland and Iceland) and a small amphidromic system enters between the Shetland Islands and Iceland, with its center not far off the Faroes.

The characteristic K_1 diurnal tide has two large amphidromic systems, one covering the entire North Atlantic and the other the South Atlantic. Its picture is simpler than that of the semidiurnal tidal waves.

For a full explanation, all the factors involved must be taken into consideration. Clearly the complicated shape of the ocean, sympathetic waves from neighboring oceans, the gyroscopic effect, and friction all have a great influence on these tides. Proudman and Doodson have drawn a purely hypothetical picture of an ocean of constant depth with meridians 180° apart. An application of these theoretical results to actual oceans is difficult, but they nevertheless indicate what sort of tides can be expected.

It is not surprising that so much attention has been paid to tides in the Atlantic, with its fairly simple shape. Stretching between the great Antarctic water basins and the Arctic Ocean, it is cut off from more northerly waters by barriers between Greenland and Iceland and between Iceland and Scotland. The tides in the Atlantic Ocean may thus be considered to be a superposition of lengthwise (longitudinal) and crosswise oscillations. While the longitudinal oscillations are made up of the independent tide and the sympathetic tide due to the Antarctic waters, the crosswise oscillations are due essentially to the gyroscopic effect (Defant, 1924). Observations have shown that such an approach is essentially justified, even though it gives only a very crude account of the natural facts. But from what we know about the formation of tides in wide seas (for instance, the North Sea) we can assume that the Atlantic tidal wave is a progressive S–N wave which, after running the length of the Atlantic basin, loses a great deal of energy in crossing the barriers separating it from the Arctic Ocean, and later in crossing the ice of the polar region. The reflected tidal wave is therefore weaker than the incident wave. The resonance tide, instead of having the form of a simple stationary wave, must be thought of as a set of superposed waves that are out of phase. The tidal wave in the South Atlantic is almost completely the progressive wave entering from the south. The Atlantic would therefore seem to be similar to the North Sea, where, as we have seen, the incident wave from the north loses so much energy due to friction in the south that the reflected wave, while still being strong enough to produce a rotatory tide off Heligoland, is so weak in the northern North Sea that the incoming, progressive wave is almost unimpeded (Defant, 1928).

This theory can be put to the test along the central E–W axis of the ocean. Here the gyroscopic effect of the rotation ought to be negligible, and also there are many oceanic islands in these parts where observations can be made and compared with theoretical results. Figure 53 shows the distribution of the amplitude (top) and phase (bottom) of the semidiurnal tide along this axis of the

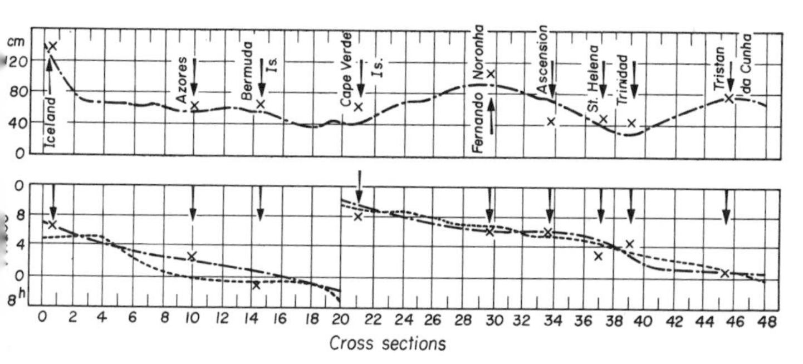

FIG. 53. Distribution of amplitude (cm) and phase (lunar hours) of the semidiurnal spring tide along the central axis of the Atlantic. Continuous curve: computed values; crosses: observed values shown in Table 6.

Atlantic Ocean, computed from theoretical considerations. The crosses are the actual values observed at the oceanic islands (see Table 6). Clearly the observed and theoretical values are in close agreement. Theoretically the semidiurnal tides along the central axis ought to be a mixture of independent and sympathetic tides, the latter being predominant. However the resulting tidal picture comes about mainly through the great loss in energy undergone by the tidal wave when it reaches the Arctic barriers.

A new method for obtaining a picture of the tides on the open ocean has recently been devised by Hansen (1949). Hansen assumes that once the coastal constants are known, the tide can be predicted for any point on the ocean. This method requires a great deal of laborious calculation, since a narrow grid system covering the entire ocean must be set up. Applied to the Atlantic Ocean,

TABLE 6.

Values of the Semidiurnal Tides on Oceanic Islands near the Central Axis of the Atlantic

PLACE	LAT.	LONG.	AMPLITUDE IN CM	PHASE. LUNAR HOURS IN G.M.T.
Heimaey (Iceland)	63.4° N	20.3° W	137	6.5^h
Western Azores (average of 8 stations)	38.7° N	28.5°	60	2.1
Bermuda	32.3° N	64.8°	55	11.2
Cape Verde Is. (2 western stations)	17.0° N	25.2°	63	8.0
Fernando Po, Brazil	3.8° S	33.1°	105	6.6
Ascension I.	7.9° S	14.4°	45	6.3
St. Helena I.	15.9° S	5.7°	45	3.5
Trinidad, Brazil	20.5° S	29.1°	42	5.5
Tristan da Cunha	37.0° S	12.3°	75	0.8

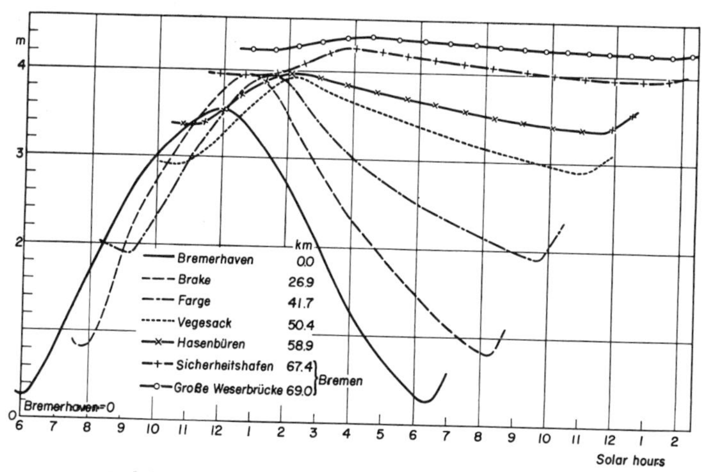

FIG. 54. Tidal curves in the lower Weser (1879) before the river was extended. (After Franzius.)

the method has shown good results, but it does not attempt to establish the origin of the tidal picture.

Tides in River Estuaries

Oceanic tides produce oscillations upriver in estuaries open to the sea. The Amazon is probably the river with the longest tidal stretch (more than 500 miles). The tide wave entering the river from the deep sea loses its symmetrical form due, first, to a fall in depth and the resulting greater friction, and second, to the flow of the river. These influences are of a hydraulic kind and vary with local conditions.

The greater the distance from the estuary, the more asymmetrical the tidal curve. The water level rises more and more quickly and falls more and more slowly, so that the flood tide is much briefer than the ebb tide. Figure 54 shows the tidal curves of the lower Weser before it was broadened in 1879. At the time it flowed in a broad and shallow bed between banks supported by dykes.

While the tidal range in a closed basin increases with distance from the opening, in rivers it decreases upstream, and, in the case of the Weser, it ceases completely a little above Bremen (see Table 7).

At the estuary the tide takes 6 hours to rise or fall, but near Bremen it rises for less than 3 and falls for more than 9 hours. The symmetry of the tidal curve is lost, and the oscillation is no longer regular. This increasing

TABLE 7.

Tides in the River Weser Before It Was Broadened

POINT	KM	DURATION OF				HEIGHT OF		
		WATER RISE		WATER DROP		LOW WATER	HIGH WATER	RANGE
		HR.	MIN.	HR.	MIN.	M	M	M
Bremerhaven . . .	0.0	5	57	6	28	0.26	3.56	3.30
Brake	26.9	5	1	7	24	0.81	3.96	3.15
Farge	41.7	4	17	8	8	1.90	3.94	2.04
Vegesack	50.4	3	32	8	53	2.91	3.93	1.02
Hasenbüren	58.9	3	5	9	20	3.37	3.96	0.59
Bremen Sicherheitshafen	67.4	3	2	9	23	3.95	4.28	0.33
Bremen Grosse Weserbrucke	69.0	2	58	9	27	4.23	4.40	0.17

asymmetry is due to the fact that within the tidal portion, the velocity of the crest of the tide wave (high water) differs from the velocity of the trough of the wave (low water). Between Bremerhaven and Brake, for instance, the velocity of the crest is 9.5 m per sec, while that of the trough is 4.3 m per sec; between Wegesack and Bremen the figures are 2.7 m and 1.8 m per sec respectively. These differences are due mainly to friction and the flow of the river. While the former reduces the return flow, the latter increases it, so that the flow of water upriver is impeded during high tide and increased downriver during low tide. This uneven tidal flow results in a steepening of the curve of the rising tide. Near estuaries, the flow is generally greater at flood than at ebb, turning soon after high water. The interval increases with distance upstream. In the Weser there is hardly any flow at flood above Wegesack (only 0.2 m per sec), while at ebb the flow has a velocity of 0.7 m per sec.

In many rivers the profile of the incoming tide wave changes so much that we have a *bore*, during which the water surges upriver with a high, steep front. This striking phenomenon is usually associated with shallow depths and sudden increases in the gradient of the river bed. There are no bores in German rivers; but bores occur in some French rivers, for instance the Seine, Orne, and Gironde. The bores in the rivers Severn and Trent (England) are well known. Tremendous bores occur in the river Petitcodiac at the northern end of the Bay of Fundy, New Brunswick, Canada, and particularly in the Amazon River, where this phenomenon is called the Pororoca. The Pororoca is often so great as to make the river impassable. From the banks it looks like a mile-long waterfall, up to sixteen feet high and travelling upriver with a speed of twelve knots. Its roar can be heard almost fifteen miles away. Figure 55 shows a picture of the bore on the river Fuchun (Tsientang) in

China, where it is said to have reached a height of up to twenty-five feet. Clearly increases in the asymmetry of the wave profile of river tides must finally culminate in bores. The almost vertical wave front moves forward very quickly, continuously overtaking, as it were, the shallower waters ahead of it. The tide wave has become surflike, now resembling a progressive pulse rather than a periodic oscillation of the water level.

Similar phenomena occur in other waters besides rivers. The sudden rise of the water in the channels of mud flats is probably due to similar causes, and so, too, is the onrush of water in narrow straits with different tides on either side, e.g. the Strait of Messina between Sicily and Italy (the ancient passage between Scylla and Charybdis).

FIG. 55. Bore on the Fuchun (Tsientang) River.

VI. *Internal Waves and Tides*

The earth's water envelope is not homogeneous, since differences in temperature and salt content alter its density (specific gravity). Warmer and less salty water, being lighter than colder and more salty water, generally rises to the top. At the point of contact of two definite layers, there are usually quite sudden changes in density (transition layer). In general, density increases in direct proportion to depth, but there are frequently slight irregularities in the density gradient, which make possible oscillations of quite a different kind to the oscillations of the water surface which we have discussed so far. Here the maximum displacement takes place at the point of contact between two strata of water of different density, the surface itself remaining almost smooth. For this very reason these oscillations have escaped investigation until fairly recently—they cannot be detected from above, but can only be demonstrated by a rapid succession of measurements of temperature and salinity at depths having a maximum density gradient.

Internal waves in the ocean were first demonstrated by measurements carried out from a ship at anchor. These waves are usually affected by meteorological factors, and often have the same period as the tide, so that they are probably influenced by tide-generating forces also. Otto Pettersson was probably the first to demonstrate internal waves with the same period as the semidiurnal tide in his work on the flow of water between the Baltic and the Kattegat. He even discovered a way of obtaining a direct record of the internal tides: he recorded the vertical displacement of a sinker suspended in the transition layer. Figure 56 gives such a record over five days. The amplitude of the semidiurnal waves is occasionally greater than five meters, e.g. many times that of the surface tide, which, in the harbors of the Kattegat, rarely exceeds 30 cm. On the open sea, special ships are needed for recording the short-period fluctuations in temperature and salinity associated with internal waves. The "Meteor" expedition (1924–27), particularly, managed to gather such observational material, and Figure 57 shows conditions recorded at Station No. 254 in the Atlantic (2°27′ S., 34°57′ W.) where a series of twenty-three readings down to 200 m were taken at two-hour intervals. A well-developed transition layer was discovered at a depth of 100 m. An almost homogeneous layer of light water covered an equally homogeneous layer of heavy water. The figure shows changes in temperature at various depths. Apart from one sudden jump (at midnight on January 31), a wave of period 12.3 hours oscillates about the transition layer (at depth 100 m). Clearly this is the period of the semidiurnal lunar tide. The layers above and below the transition layer show hardly any

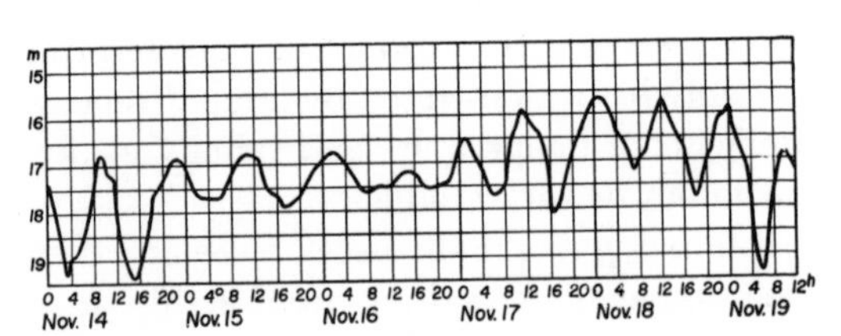

FIG. 56. Record of internal tides in the Kattegat. (After Kallenberg.)

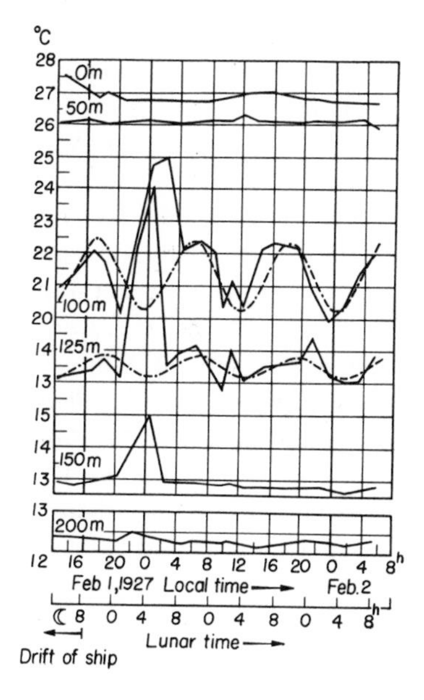

FIG. 57. Temperature fluctuations at different depths. Recorded at 2°27′ S, 34°57′ W, on anchor-station "Meteor" No. 254 (January 31 to February 2, 1927). The broken curve is the main lunar constituent tide.

oscillations and this is equally the case for fluctuations in salinity. Figure 57 must then be looked upon as the graph of an internal tide wave with a mean amplitude of 3.6 m and a phase of 4.3 lunar hours (after the moon's passage through the Greenwich meridian).

Waves near the transition layer are rather easily excited, since only small differences in density have to be overcome. Waves at the surface of water require much more energy for their excitation, since the change in density from water to air is very much greater. Furthermore, the velocity of propagation along the transition layer is very small compared with that at the surface, so that along this layer the waves require much more time to cover a given distance. J. E. Fjeldstad measured changes in temperature and salinity for 88 hours, together with simultaneous measurements of flow, at a number of stations in a fiord, and found that the velocity of propagation of the longest wave was 63 cm/sec and that the whole system of these internal waves was so regular that a theoretical interpretation was possible. There is no doubt that all oceans have internal waves of a tidal character and that these are generated and propagated according to special laws. We still know very little about the internal tides of the oceans, observations being inherently difficult, laborious, and costly.

When a fairly large oceanic area of given temperature and salinity is disturbed by external factors (e.g. quick changes in air pressure or direction of surface wind), it will tend to return to equilibrium position. This return takes the form of fairly large oscillations about the final position of equilibrium. The amplitude of these oscillations will depend on the strength of the external disturbance; but we know (see p. 61) that their period is determined by the characteristic dimensions of the oscillating system, namely the oceanic region affected by the disturbance. If the particular oceanic region has a well-developed transition layer, the internal oscillations may last for a very long time and are only damped by friction. Now this whole process happens to take place on a planet that rotates about its axis once every 24 hours. The gyroscopic effect of the earth's rotation (Coriolis force) comes into play with consequent alterations in the period of the internal oscillation. If the oceanic region is large enough, the period of this new effect will be largely due to its mass (inertia) and will vary with the Coriolis force. The period of such an oscillation is 12 hours at the poles, and increases considerably towards equator (60° latitude: 13.9 hours; 30° latitude: 24 hours; 5° latitude: 138 hours).

We ought thus to expect, besides the internal waves already discussed, the occurrence of internal inertial oscillations about the transition layer. Thus the whole picture becomes very confused. Occasionally the data obtained on ships at anchor are sufficient for an understanding of the individual factors involved. During the international investigation of the Gulf Stream (1938), investigators on the German ship "Altair" spent almost four days taking readings of temperature and salinity and of the corresponding rate of flow down to a depth of 1,000 m. The observations clearly proved the presence of a semidiurnal tide wave in the well-developed transition layer, and also the presence of another oscillation with a period of roughly 17 hours, which happens to be that of the inertial oscillation for the particular geographical latitude of the observational point (44°33′ N). The fluctuations in temperature and salinity must be the result of the superposition of the two waves, and this is in fact what is found. Figure 58 shows the mean temperature changes in a layer of between 25 and 75 m depth (the transition layer) during the period of the observations. The two thin curves represent the semidiurnal tide wave (period 12.3 hours) and the inertial wave (period 17.1 hours) respectively, and the thick curve their superposition. The dots represent the current measured by instruments after smoothing of irregularities. The curves are almost in complete agreement and there can hardly be better proof of the periodic behavior of internal waves.

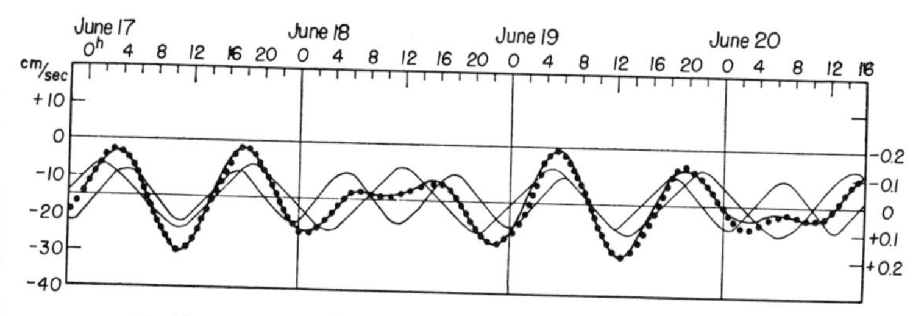

FIG. 58. Temperature fluctuations in a layer between depths 25 m and 75 m, determined from the superposition (thick curve) of the inertial and semidiurnal oscillations (thin curves). (The temperature scale has been reversed.) The dots represent the east component of the measured current after irregularities were smoothed out.

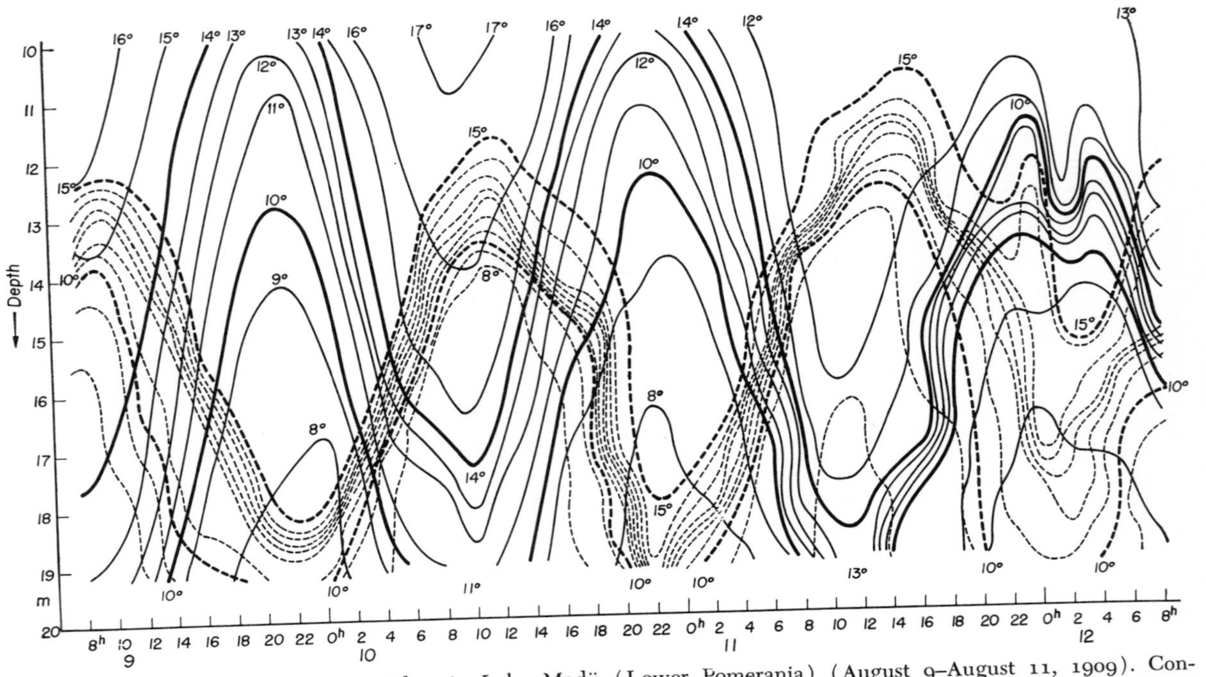

FIG. 59. Internal temperature seiches in Lake Madü (Lower Pomerania) (August 9–August 11, 1909). Continuous curves: isothermal lines at a station in the north of the lake; broken curves: isothermal lines at a station in the south of the lake.

In completely or partially enclosed basins these internal oscillations are even more marked since, in lakes, for instance, the water is highly stratified in the summer with a well-developed transition layer. Such internal waves are then stationary, i.e. they are internal seiches, and they, too, can only be demonstrated by successive temperature measurements at different depths. Again, we have little observational material to go by. Figure 59 is a record of a temperature seiche observed by Halbfass and Wedderburn in Lake Madü in Lower Pomerania, a long narrow lake. A very clear transition layer was situated at a depth of 15 m. Fluctuations in temperature are represented by isothermal lines. The continuous and broken curves refer respectively to temperature changes observed at the northern and southern ends of the lake. The two sets of curves are completely out of phase. The seiche is mononodal, and has an amplitude of roughly 6 m and a period of almost exactly 25 hours. The period can be computed from the relative thicknesses of upper and lower layers, and the computed value agrees with the observed data. Free internal oscillations take place also in larger enclosed seas, fiords, bays, and channels, but there are too few observations for a detailed picture to have emerged.

VII. *Tidal Oscillations in the Atmosphere and in the Ionosphere*

General

The tide-generating forces of moon and sun act also on the air and produce atmospheric tides in it. In 1774, Laplace applied his dynamical theory to these tides as well. He predicted that they would be very small, despite the common belief that moon, sun, and all the other celestial bodies ought to have a far greater pull on the far more elastic masses of the atmosphere than on the less responsive oceans. Clearly this is not the case, since, as we have seen, the tide-generating forces vary with the mass of the attracted body. We have also seen unit volume of water is a thousand times as heavy as unit volume of air. Assuming equal elasticity in both substances, atmospheric tides would be a very small fraction of oceanic tides, in fact, the atmospheric lunar tide would cause a maximum pressure change of $\frac{1}{27}$ mm Hg (atmospheric pressure is measured by the height of a column of mercury). The concentration of air would be strongest wherever the moon was at its zenith or nadir (0^h and 12^h lunar time). At these times there would be "high tide" in the atmosphere, while at 6^h and 18^h there would be "low tide."

Now there is in fact a daily cycle in the air pressure

with a semidiurnal constituent (a tide having a period of 12 solar hours). In the tropics, where nonperiodic barometer fluctuations are small and insignificant, the barograph records a regular semidiurnal wave with pressure maxima at 10 a.m. and 10 p.m., and pressure minima at 4 a.m. and 4 p.m. Here the fluctuation is roughly 2 mm Hg. This fluctuation decreases with higher geographic latitude, and in temperate latitudes these semidiurnal oscillations in air pressure are completely obscured by large nonperiodic changes arising out of the movements of high and low pressure areas which are characteristic of middle and high latitudes. In these latitudes, the semidiurnal pressure wave can only be isolated by laborious observations and calculations. Von Hann, particularly, has been able to show the periodic nature of this pressure wave. There is also a diurnal air pressure wave, but this wave is very irregular both in phase and in amplitude, and depends very much on local conditions.

Now these semidiurnal pressure fluctuations cannot possibly be the atmospheric counterpart of ocean tides, since they follow solar and not lunar time. If there were a solar tide, there would have to be an even stronger lunar tide, which is not the case. The semidiurnal wave is probably caused by semidiurnal temperature changes, which produce greater effects than the tide-generating forces. The possible tidal effects are completely obscured by the far greater thermal effects.

The moon on the other hand does not affect the temperature of the atmosphere, and it is thus possible to isolate the direct tide-generating effect of the moon on the air. These lunar atmospheric tides are very small and a great many observations are needed to exclude what are generally large, nonperiodic fluctuations in air pressure. This tremendous numerical work was undertaken particularly by J. Bartels and S. Chapman, who managed to isolate pure lunar tides at a number of places. Bartels analyzed 150,000 hourly air pressure readings taken in Potsdam and Hamburg for each of 66 years to demonstrate lunar oscillations of a little more than 0.01 mm Hg for these places. In tropical regions the fluctuation is somewhat larger; Figure 60 shows a tide in the air pressure at Batavia over 12 lunar hours, calculated from observations made over forty years. The amplitude of 0.064 mm Hg is roughly six times that found in our latitudes. Air pressure maxima occur at 0.8 and 12.8 lunar hours, which is not long after the passage of the moon through the meridian. This wave can be represented by a period clock (Fig. 60, right): a hand (vector) being drawn from the center to a circumference graduated according to the moon's time. The vector points to the time of maximum, its length gives the amplitude, or greatest deviation from average air pressure. Despite the very small amplitude of this wave, the regularity of the phenomenon is striking. Similar results for different points have been obtained by other investigators. The maximum always coincides fairly well with the moon's transit through the meridian; nowhere are the deviations greater than an hour. The amplitudes are greatest near the equator and decrease towards the poles. Figure 61

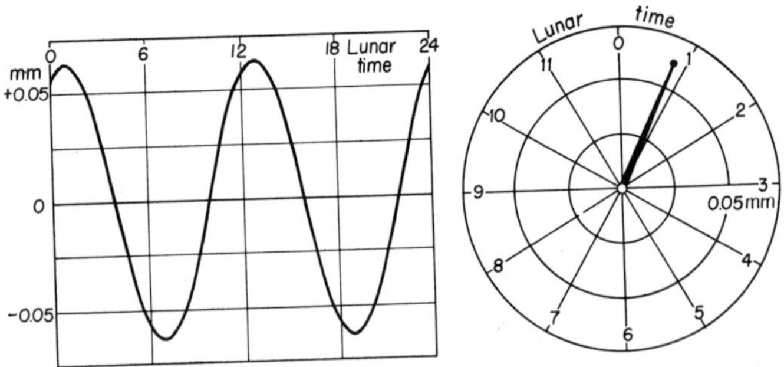

FIG. 60. Semidiurnal air pressure wave in Batavia (computed from observations over 40 years). Right: its representation by a period clock. (After J. Bartels.)

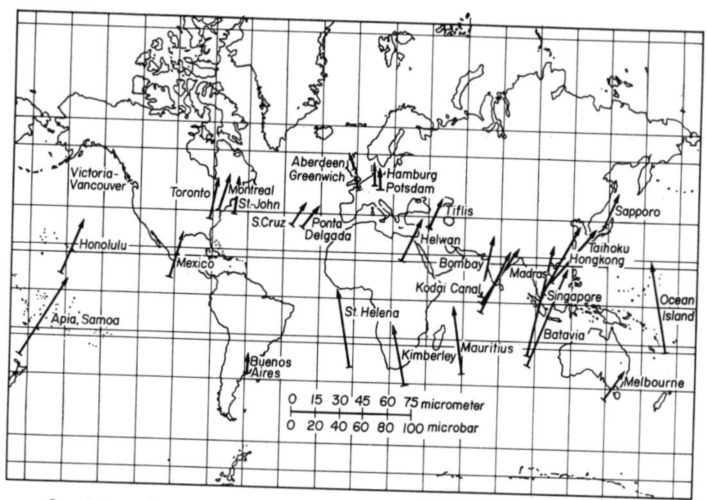

FIG. 61. Mean lunar air pressure wave for 31 points represented by the vectors of a period clock. The center of the vectors shows the location of the point; their length is proportional to the amplitude. Singapore has the greatest amplitude with 0.067 mm Hg. A vertical arrow pointing north would mean that the crest of the pressure wave coincides with the moon's transit through the given meridian. (After Chapman.)

shows the air pressure wave during 12 lunar hours for 31 points, represented by the vectors of a period clock. The center of each hand indicates the position of the point considered. Singapore has the greatest amplitude with 0.067 mm Hg. The wave usually reaches a maximum after the moon's transit through the meridian, and within an interval of less than an hour. The wave varies with distance from the moon, but there are also unexplained seasonal variations.

Thus there is no doubt that there are atmospheric lunar tides, and we must therefore assume that there are bound to be atmospheric tides due to the sun as well. Unfortunately these tides must needs have the same period as very much greater temperature fluctuations. To sum up, the effect of the tide-generating forces on the atmosphere appears as a relatively simple and regular pressure wave travelling from east to west, which can only be measured by means of sensitive instruments. The reason why atmospheric tides are so much simpler than oceanic tides is mainly due to the fact that the atmospheric envelope covers the entire earth and that there are no barriers to their free movement, such as the continents are to the movement of the waters of the earth.

Tides in the Ionosphere

There are tides even in the ionosphere, which starts some 60 miles above the surface of the earth, and these may well be greater than those in the lower atmosphere. Although direct observations in these high layers involve great difficulties, the effect on terrestrial magnetic phenomena can be investigated. The sun ionizes these layers, and displacements of ionized air due to the tide-generating force produce periodic electric currents in these high layers with a consequent disturbance of the magnetic field of the earth. Such fluctuations are known as magnetic tides.

Terrestrial magnetic tides have clear solar and lunar components, just like the oceanic tides. Figure 62 shows the residual lunar magnetic oscillation after deduction of the solar magnetic constituent at Greenwich and Batavia over one lunar day. Clearly the curve is a regular one.

From these magnetic variations we can deduce the strength of the electric currents in the ionosphere. The system earth-atmosphere-ionosphere is looked upon as a gigantic generator with the earth as its permanent magnet. The atmosphere acts as the coil, set in motion (a) by thermal differences caused by solar rays and heat losses through radiation, and (b) by the tide-generating forces of moon and sun. These thermal and tidal movements are like a "breathing" of the atmosphere. The electrically conducting layers of the ionosphere may be thought of as the windings of the coil. Since these move perpendicular to the earth's magnetic field, induced currents are produced, whose magnetic effects can be observed on earth. The exact analysis of these magnetic fluctuations is very fruitful and shows particularly well that there are small systematic fluctuations associated with the position of the moon (see Fig. 62). Since the light from the moon has no thermal or ionizing effects, the associated displacements of air in the atmosphere must be due to its tide-generating force. The magnitude of the luni-periodic electric currents of the ionosphere and thus the lunar variations of terrestrial magnetism depend only on the degree of electrical conductivity of the ionospheric layers, which are caused by ionization through solar radiation. Thus magnetic tides are a fair indication of the degree of ionization in the higher strata of the air caused through ultraviolet radiation. Since this radiation increases a great deal between sunspot minima and sunspot maxima, the solar and lunar terrestrial magnetic tides must fluctuate accordingly. Observations show that this is in fact the case. These relations are one of the most important sources for our understanding of the physics of the higher atmosphere.

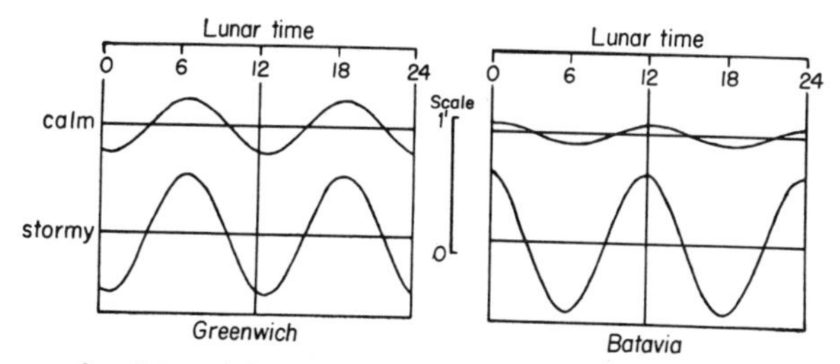

FIG. 62. Diurnal fluctuation in the magnetic declinations in Greenwich and Batavia on (magnetically) calm days (top), and stormy days (bottom). The fact that the waves for the two stations are reversed follows from their respective situations in the Northern and Southern hemispheres. (After J. Bartels.)

VIII. *Tides on Land*

In our discussion in Chapter III of the horizontal pendulum and the gravimeter, we pointed out that measurements of the tide-generating force clearly proved the existence of tides on land. On a completely rigid earth, the instruments would be standing on an immovable plane and all readings would correspond to the theoretical values of the tide-generating force. On the other hand, if the earth itself responded fully to the tide-generating force the instrument would record no changes whatsoever. Now the maxima recorded by the instruments are generally smaller than the theoretical maxima and there are also phase differences. These discrepancies show that the earth is neither rigid nor fully elastic.

We have seen that in Marburg and Freiberg the horizontal deflection (see pages 35 and 37) was only two-thirds of the theoretical value, and that similar differences resulted from gravitational measurements in Marburg and Berchtesgaden. Table 8 gives some more recent observations: the most probable value of the ratio between the observed deflection and the theoretical deflection is 0.69. Thus the earth's crust is deformed by the

TABLE 8.

Results of Observations of Variations in Deflection

INSTRUMENT	OBSERVER	PLACE	YEAR	TIDAL CONSTITUENT	COMPONENT	RATIO OF OBSERVED TO THEORETICAL DEFLECTION
Horizontal Pendulum	Schweydar	Freiberg	1910–1915	M_2	North	0.54
				M_2	East	0.61
Horizontal Pendulum	Schaffernicht	Marburg	1934	M_2	North	0.65
				M_2	East	0.87
Horizontal Pendulum	Gnass	Berchtesgaden	1937	M_2	North	0.53
				M_2	East	0.74
Double (Horizontal) Pendulum	Lettau	Collm-Leipzig	1936	M_2	East	0.58
		Berchtesgaden	1938	M_2	North	0.40
Variometer	Egedal and Fjeldstad	Bergen (Norway)	1934	M_2	North-East	0.58

tide-generating forces, and the above ratio is a measure of its relative rigidity. This rigidity is of the same order of magnitude as steel, so that we may assume the earth to respond to the tide-generating forces as if it were a ball of steel of roughly the same size.

According to Tomaschek's measurements in Marburg, the M_2 land tide has an amplitude of roughly .5 m, i.e. the earth's surface rises and falls twice daily by .5 m (20 inches). Naturally this rise is imperceptible, as there is no fixed point from which to measure the fluctuations, since the observer standing on the earth undergoes the same motion.

More recently Eckhardt, using an extremely accurate gravimeter, determined simultaneous tidal fluctuations in the value of gravity at different latitudes. Figure 63 shows his observations at a number of different points over four consecutive days. While the observed fluctuations are fairly regular, they are generally greater than the theoretical values.

There is one other method for determining the deformation of the earth's crust by the tidal forces. Oceanic tides are generally assumed to take place over a rigid sea bed, i.e. over a rigid earth. Now if the sea bed itself rises and falls under the tidal force, our water gauges would no longer be recording the true fluctuations in sea level but rather the differences between these and the vertical displacements of the sea bed. If we could compute the true oceanic tides for a coastal point, then the difference between the computed range and the observed range would be due to the simultaneous vertical displacement of the sea bed caused by tides in the rigid crust of the earth. Now, oceanic tidal phenomena are generally so complex that it seems impossible to compute theoretical tides for given coastal points with sufficient accuracy. However, some tidal constituents lend themselves to such a computation, for example, tides with a long period. Of these, the fortnightly lunar tide M_f is by far the most important (see Table 1, page 48). Its period is so long that it actually obeys equilibrium theory and "tidal mountains" have time to be formed (see p. 41). The theoretical tidal range can then be calculated very accurately for the case of an absolutely rigid earth. A comparison between these computed values and the values obtained from the harmonic analysis of observational data then leads to values of the land tides. Thompson, taking readings at selected oceanic points over many years, obtained an average ratio between the

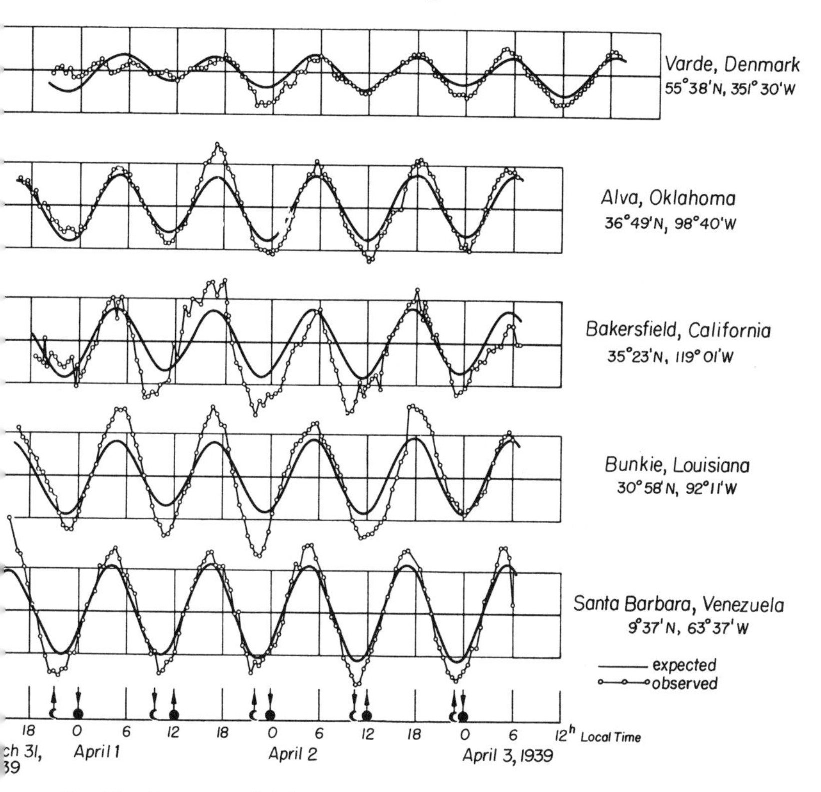

FIG. 63. Simultaneous tidal fluctuations in the acceleration due to gravity at 5 points (March 31–April 3, 1939). (After E. A. Eckhardt.)

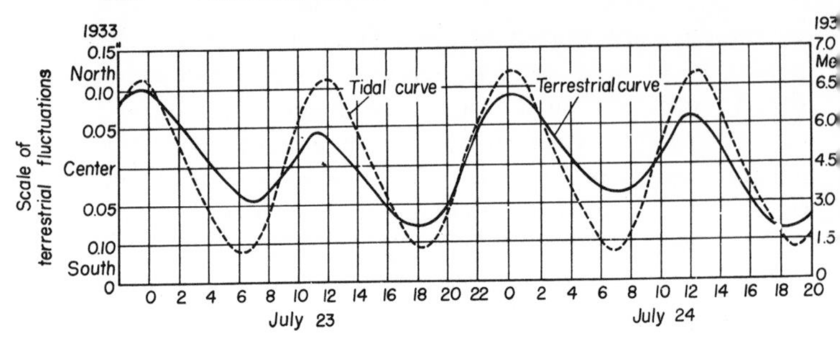

FIG. 64. Relation between the tide off Liverpool and fluctuations of the land surface at Bidston (after Doodson and Corkan). ---------: tide; ———: land fluctuation.

observed and the computed tidal range of 0.68 for the M_f wave, while Schweyder, using even more data, obtained 0.66. These values are in close agreement with the results obtained by means of the horizontal pendulum. More recently Proudman and Grace, applying hydrodynamic theory to the M_2 tide in enclosed seas (Red Sea, Gulf of Suez) and large lakes (Lake Baikal), showed that fluctuations of the sea bed (earth's crust) could be deduced from the ratio between the observed and theoretical values of even this tidal constituent. There have been only a few observations but these have given values in the same range: Red Sea, 0.75; Lake Baikal, 0.73; Gulf of Suez, 0.67.

Because of recent great improvements in instrumental techniques, still further components in the fluctuations of the earth's crust have been demonstrated. The chief of these is a diurnal fluctuation along the east-west plane, which is most probably due to daily temperature differences. There are also annual and secular fluctuations, and clearly lengthy analyses are needed if a clear picture of the tides in the earth's crust is to emerge. Furthermore, the apparently rigid crust also responds to stresses directly on it. Such stresses are particularly great near the coast. During ebb and flood, millions of tons of water are alternately withdrawn and thrown on to a particular coastal region. Not only do the changes in the mass of water during high and low tides produce corresponding gravitational effects over the coast, but they clearly subject the bed to constant pressure changes. All these fluctuations are faithfully recorded by the horizontal pendulum, thus obscuring the actual tides of the rigid crust itself, particularly since all these fluctuations must needs have the same period as the tides. Observations off the coast of Japan have shown that the fluctuation of the earth's crust due to changes in the water mass is more than fifty times as great as that caused by the tides. In Bergen, Norway, both effects were found to be of the same order of magnitude, and even in Freiberg, Germany, there was a perceptible change in stress. Figure 64 clearly illustrates the close connection between oceanic tides and fluctuations in the sea bed.

There are a whole host of other factors which can cause periodic or nonperiodic fluctuations of the earth's crust. The earth is covered by an atmosphere that exerts different pressures upon it as the barometer rises or falls. These pressure effects are fairly large, and when the barometer fluctuates by 25 mm Hg the pressure difference on each square meter of earth is 340 kilograms (approximately ½ lb/sq in). Similar effects are also due to considerable concentrations of snow (e.g. on mountains in the winter) and to considerable downpours within a limited area, etc. As observations at strategic points increase in number, so are we likely to obtain a greater insight into the problems of land tides and the deformations of the earth's crust.

INDEX

THE ANTS by Wilhelm Goetsch

THE STARS by W. Kruse and W. Dieckvoss

LIGHT: VISIBLE AND INVISIBLE by Eduard Ruechardt

THE SENSES by Wolfgang von Buddenbrock

EBB AND FLOW: THE TIDES OF EARTH, AIR, AND WATER by Albert Defant

THE BIRDS by Oskar and Katharina Heinroth

PLANET EARTH by Karl Stumpff

ANIMAL CAMOUFLAGE by Adolf Portmann

THE SUN by Karl Kiepenheuer

VIRUS by Wolfhard Weidel

THE EVOLUTION OF MAN by G. H. R. von Koenigswald

SPACE CHEMISTRY by Paul W. Merrill

THE LANGUAGE OF MATHEMATICS by M. Evans Munroe

CRUSTACEANS by Waldo L. Schmitt

TWO-PERSON GAME THEORY: THE ESSENTIAL IDEAS by Anatol Rapoport

THE UNIVERSITY OF MICHIGAN PRESS